CRACKING the SKY

CRACKING *the* SKY

A HISTORY OF ROCKET SCIENCE IN SOUTH AFRICA

Desmond Prout-Jones

University of South Africa
Pretoria

© 2002 University of South Africa

ISBN 1-86888-203-9 (soft cover)

Published by Unisa Press
University of South Africa
PO Box 392, 0003 Pretoria

Cover design and layout: Jörg Ludwig
Typeset by Jörg Ludwig
Printed by Interpak, Pietermaritzburg

© All rights reserved. No part of this publication may be reproduced in any form or by any means – mechanical or electronic, including recordings or tape recording and photocopying – without the prior permission of the publisher, excluding fair quotations for purposes of research or review.

Contents

Foreword — vii
Preface — x
Acknowledgements — xii
Author's note — xiv
Introduction — 1
Chapter 1
   The fuse is ignited — 4
Chapter 2
   Reaching for the elusive mile — 22
Chapter 3
   Of mice and rockets — 35
Chapter 4
   The fire that thunders — 51
Chapter 5
   Events with a double bang — 64
Chapter 6
   Mighty flights by courtesy of Prometheus — 76
Chapter 7
   The final countdowns begin — 86
Photographic section
   The photo gallery — 93
Chapter 8
   Even the gods cheered — 131
Chapter 9
   Lifted on tongues of fire — 143
Chapter 10
   A look at history: fire that moves — 149
Chapter 11
   Fire that moves: the rocket — 159
Chapter 12
   The last chapter — 184
Epilogue
   The fire is no more — 189
References — 192

This is for my friends and colleagues of the South African Rocket Research Group (SARRG), who assisted me and believed in me.

Foreword

For thousands of years the stars have held a fascination for humankind, who has been drawn to reach them. The rocket holds the key to the exploration of space. During World War II the V-2 rocket was developed for military purposes because it could be guided to a target. In combination with other rocket stages the V-2 proved that space travel could become a reality. It was soon after the war that a few groups of amateurs around the world, including two groups in South Africa, determined that it was feasible to make a rocket that could go into space and possibly even place a satellite in orbit. Their enthusiasm and creativity were infectious.

The two world powers, the USA and the USSR, entered an unwritten competition to be the first to conquer space. This race became a national competition and a serious matter of pride. This added to a general level of interest in rocketry and space travel that permeated the rest of the world, including South Africa. This was part of the spark that triggered the local interest in amateur rocketry. It became the burning desire of a few rocketeers to push the limits of this science and to do for rocketry what other amateurs were doing for radio.

The two active local rocket research groups working independently of each other followed almost identical paths in their research, though the South African Rocket Research Group (SARRG), led by Des Prout-Jones, had begun their experiments several years in advance of the other group. It is uncanny to see the designs, successes, failures, and solutions to problems of the two groups. They are almost identical in their major facets.

Both groups determined to make a rocket that would go into space. The similarity in the targets and goals that they set for themselves was based on the common vision they had. A common vision will always lead along similar paths to the ultimate goal.

The results that were independently achieved were impressive. Creativity, innovation and irrepressible enthusiasm enabled these teams to achieve successes that are impressive by any standard even today. It is a pity that the two groups could not have worked together as a single team. Time and distance had made this virtually impossible. Unfortunately, the hopes and aspirations of both groups met a similar fate at the hands of the same person in the seat of bureaucratic authority.

Records were set more than 50 years ago that in some cases hold to this day. As the groups worked towards their goals there was always the sword of Damocles hovering above their heads. In this case, the Chief Inspector of Explosives. It is against this background that rocketry began in South Africa. On the one hand there were a handful of amateurs with an unbelievable dedication to realising their dreams. On the other was a hostile bureaucracy and sections of the public. It was these teenagers and young adults that against all odds laid the foundations of rocketry in South Africa. They had only one goal in mind and that was to do everything in their power to tame the rocket for scientific use and consequently develop a space vehicle.

There is an unbelievable thrill that comes from pushing the frontiers of knowledge, especially when it involves controlling the awesome power contained in a small rocket motor. Each advance was only a small creative step forward, but each of these steps was also a learning experience that has never been forgotten by the people that were involved in taking them during the mid-nineteen-hundreds in South Africa. We must include the support that these pioneers got from their families, especially when things went wrong.

The story of rocketry in South Africa is a thrilling, fascinating account of how creativity can be nurtured by supportive people. It is also a story about how a potentially valuable creative national resource was destroyed by authority. It proves that the creative human mind is potentially the most valuable resource that any country can have. We must not suppress creativity, but must nurture it to the benefit of humankind. We must also be vigilant about how we treat our creative human resources.

It is when we allow people to dream that the things that were once thought to be impossible become a reality. The same engineer, Wernher von Braun, who made the infamous V-2 rocket, was responsible for the Saturn moon-landing vehicle. When ideas become too big for authorities, they suppress them, but when ideas grow too big for one creative person that person builds a creative team to achieve his or her goals.

Only a few of these intrepid South African rocket pioneers are alive today. They still dream about the satellite that might have been. The photographs are still tucked away in their desks and pasted into scrapbooks. This book is only one part of the fascinating history of the early days of amateur rocketry in South Africa dating from around 1947 and ending abruptly in 1962. Enjoy reading the book, but never forget the underlying lesson of what can be lost when we fear the future and suppress the creativity that opens up the blue skies for us.

Prof Cedric Smith
Technikon Pretoria 2002

Preface

I first met Des Prout-Jones towards the end of 2000 when he was invited to address The Club on the topic of early rocketry in South Africa. The Club is a Benoni-based speakers' forum that has been in existence since 1994. The audience were so intrigued with what they heard on that Wednesday evening that the members resolved that Des should be invited again to finish the 'whole story'. This Des undertook early in 2001 and he once again held the audience spellbound when he related the experiences of the South African Rocket Research Group (SARRG) and their launches in those early years. After the lecture, several club members implored Des to publish and thereby register the history of early rocket launches in South Africa and, more specifically, his leading role in the enterprise.

All the members of The Club and I were delighted to learn that Unisa Press had decided to publish *Cracking the sky*. Des Prout-Jones's account of what was achieved so many years ago is an important chapter of technological accomplishment that is now being laid into the public domain. It could so easily have been lost forever!

The book makes for exciting reading, even for those of us who are not rocket scientists. Des relates his passion for the science of rockets with a contagious enthusiasm. The text is not cluttered with technical details and formulae and yet Des succeeds in conveying the basic principles involved in rocketry in a lucid and clear manner. The harnessing of the awesome power of chemical and liquid propellants to launch vehicles towards the limits of space is not for the faint-hearted or the foolish. Reading the text one almost becomes impatient with one's own excitement for the next countdown to learn what the more powerful launch vehicles could accomplish. Would they go higher and higher, and where would all this lead to? Sadly, having read about the shortsightedness of the authorities who stopped this venture with threats of imprisonment the reader is left with the burning question of 'Where could this have led to?' This we shall never know!

I am honoured to have been asked by Des to write a few lines as a preface to this account of events that took place on South African soil such a long time ago. What Des and his team accomplished is testimony to what can be achieved in the field of human endeavour if there is intellectual focus, dedication, commitment and passion to succeed. This book not only records, it also serves as an inspiration to what can be achieved. The Club members and I are proud to have been involved in it in a very small measure through our encouragement.

Prof Murray Faure
Pretoria
1 February 2002

Acknowledgements

It is with humility that I put pen to paper, as it were, to record the events that transformed my life at a young and impres-sionable age. As I grew older, I had the privilege of meeting many people who were sympathetic to my odd involvement in something that they themselves did not understand. Yet they encouraged me to carry on searching for the illusions that I had and even went so far as to offer suggestions and assistance in my endeavours. These people are no more, for time has passed so quickly that now my generation are the 'old timers'. This special thanks goes out to all those who had faith in what I was striving to accomplish and perhaps shared that very special dream that I have, but did not live long enough to see the final rocket crack the sky. *In memoriam*, to them I say a humble 'Thank you'.

To my five sponsors, who still wish to remain anonymous, a very profound appreciation for having the courage to stand by me when, at times, things appeared to be coming adrift. You all know who you are and my best wishes for the continued success of your businesses.

Most of the photographs in this work are the excellent expertise of my late and dear friend Jack Holloway. Apart from the great pictures, many of the intricate components are due to his competence with a lathe, and without him and his sense of perfection, I doubt whether the rockets would have left the pad!

To the members of The Club who sowed the seeds for me to write this book, my sincere appreciation, especially to Des Smith, who was the chief instigator. I had received an invitation to address this esteemed body of academics on the subject of rockets and was invited back a second time to expound further. I am grateful that I was able to answer the questions that Professor Faure asked at this second lecture, as they really put me to the test. Thank you, Prof!

My gratitude to Dr Claire Flannagan of the Johannesburg Planetarium for her unhurried explanations of things heavenly and for her valued criticism.

A very special thanks to Ms Sharon Boshoff for her patience and assistance in putting up with my faxes and telephone calls.

As always, the first come last and to my wife, Pat, goes my profound appreciation for her understanding that she had not taken a 'normal' man as her husband! She has to be a very special lady to have put up with all the strange things that occupied me throughout our marriage. For transforming a computer disk into the final manuscript, my daughter Diane stepped up to the plate and did wonders. My daughter Debra for her comments and reading the draft as it progressed. My son, Derek, for his support in whatever I became involved in and his interest in his Dad's interests.

Finally for my grandchildren, who have shown an avid affiliation to my past and are able to participate in the launching of model rockets. It makes me humble to hear their voices calling out those familiar words '… 3, 2, 1. Ignition! Lift-off!'

Author's note

The book is intended to inform the general public as well as the South African historian about the efforts made by myself in the field of rocket science. I was the first South African to design, build and launch rockets for scientific reasons. The time frame extends from 1947, when I was 11 years old, to 1963 when I was prohibited, by the government of the day, from launching rockets. I have taken the facts from my personal diaries and press clippings to put a readable account of the more prominent activities into a vehicle that may be enjoyed by all interested parties.

I have endeavoured to bring to lay people, in simple terms, the workings of solid and liquid fuel rocket motors and the advantages and disadvantages of both. Various interested bodies have invited me to lecture on the subject and to share this knowledge with others. I have attempted to give accurate accounts of the actual launchings, to set the atmosphere of anxiety and elation experienced at the firing range.

The rockets that are referred to in this work are of my own design and therefore I must take full responsibility for their performance.

With the launching of the Russian satellite Sputnik in October 1957 the public were suddenly made aware of the possibilities of space flight. The local news media had already got to know what I was doing in 1953 and this publicity brought me into contact with the Department of Explosives. It turned out to be a rather one-sided affair that ended in 1963. However, the press took every opportunity to publish my launchings, successful or not. This brought unwanted attention to my work but I realised that the publicity may have been a hidden blessing.

I was fortunate indeed to receive five sponsors from the private sector in assisting me with material to make bigger and better rockets. The cost of producing one of the 'cheaper' solid fuel rockets was far above my means. Because of the newspapers carrying the events of the firings, my name became well known, although when I launched a mouse inside a rocket, my name was not all that popular! This was just one of the aspects that rocket propulsion was used for. I was grateful that the slanted press reviews did not affect my sponsors' attitude toward my research after the mouse débâcle.

Into the late 1950s the rockets became more sophisticated and I achieved the first rocket to attain one mile (1,6 km) in altitude. This was accomplished by the successful firing of a two-stage rocket, also the first in South Africa.

Not satisfied that solid fuel offered the optimum performance, I experimented with liquid fuels and designed the first rocket with this means of propulsion, another first in this country. However, the cost of pursuing this development proved to be too great and further attempts were abandoned. The fact that the motor worked was regarded as a significant step towards ultimately developing a satellite launch rocket.

The static firings and flight firing are described.

The advance in design of bigger solid fuel rockets and their actual launches are of singular interest in the procedures that were followed to assure safety at all times. I was instrumental in the formation of the South African Rocket Research Group (SARRG) in 1959 and all rockets fired from then onwards were attributed to the group.

Explanation is given as to the final design of the method employed to ignite successive stages in multi-stage rockets. This method was effectively used to ignite up to five stages and to fire the group's last rocket. The text deals with this aspect in detail as well as the methods used to determine the altitudes achieved. Newspaper photographs and clippings support all events recounted in this book as well as the 27 black-and-white photos that illustrate it. The year 2001 is the thirty-second anniversary of the first moon landing and it is my earnest desire to inform the South African public that we were not sitting in the background. I had received publicity in the USA about my research and it is with regret that I was prevented from reaching my goal. The government informed me that if I went ahead with the launching of the rockets that would reach an altitude of 115 mls (184 km) I would face serious consequences.

This ultimatum brought an end to 16 years of practical experience and work in a subject that today is taken for granted.

20 May 2001

Introduction

From the beginning of time people have gazed into the night sky and wondered in awe at the millions of bright lights that appear. Each night for millennia, these seemingly unchanging twinkling lights have captured the imagination. We were taught at a very young age that they were, in fact, the souls of the deceased and were there to watch over us during the hours of darkness. Obviously, a very nice tale to placate any fears that the young might feel when the sun went down. It was of some consolation to know that 'Granny' was watching over us in the dark!

But human beings are peculiar creatures and their quest to learn about all things around them brought these 'night-lights' under the microscope. The centuries passed and eventually astronomy became a well-established science. The study of the stars! Today this science has been instrumental in scientists sending robotic probes to the inner and outer planets to seek information as to their composition, size, gravitational fields and anything else that is of interest to astronomers.

These scientific journeys are only made possible because of one 'thing'. The rocket. This means of propulsion is the only method that makes travel in the vacuum of space possible. The rocket, travelling faster than a rifle bullet, can achieve the enormous speeds required to break away from the gravitational attraction of the Earth.

By utilising this propulsive force and sending fact-finding probes, not only to our neighbouring planets, but also into deep space, we may be able to uncover our very beginnings. For are we not part of the cosmos, part of the very grand nature of the universe of which Earth appears as a blue orb in the vastness of space?

This book is about a young boy's dream of one day building a rocket and sending it into the air. At first glance it seems harmless enough but as the years roll on, the dream starts to become a very real day-to-day obsession. In this recollection of my activities in the field of rocket science, I have read through old newspaper clippings and also my own diaries to ensure that what is presented to the reader is the truth. I have not changed any names, but for reasons concerning my last launch, the actual details are known only to my colleagues and myself.

The fiery beginnings go back to 1947. A terrible war had finally come to an end and I could look to a future where the news broadcasts were not filled with disaster. At the age of 11 I was an avid listener to the serials that were broadcast over the radio. (There was no TV!) My favourite programme was of a science fiction nature and I used to sit in front of the radio, taking it all in! In my mind I could visualise all that was audibly being played out. The spaceships engaged in battle with the aliens from way out in space who were trying to take over our world! It was riveting stuff! And I dared not miss an episode.

The spaceships became a reality to me and I dreamt about them almost every night.

So, from the childish dreams sprang an ambition that was to last until 1962 when the entire dream was dashed, once and for all.

In my endeavours, I had the privilege of encountering men who shared my quest to launch bigger and better rockets and were willing to give up their time to achieve these ends. Apparently I was not the only nut in this fruitcake! Books on the subject of rocketry were unheard of at that time so I learnt the hard way. The only rocket that the public were aware of was the infamous V-2 that the Germans had launched against the United Kingdom. It was a terrifying weapon as the ground explosion occurred seconds before the sound of the approaching missile was heard. This was because the speed of the V-2 was faster than the speed of sound. The designer of this rocket was Dr Wernher von Braun, whose dream was to send his rockets to other planets and not to destroy cities on our planet! (More about him later.) One V-2 that was fired straight up reached an altitude of 114 mls, and this record remained for many years until exceeded by the American Viking, which reached 135 mls.

All these exciting events took place during the late forties and early fifties and it was not until 1957 that I was able to order a book from the States on the theory of rocket design; I have it to this day and it is regarded as the definitive work.

With the acquisition of this book, I was hard pressed to understand the nomenclature as well as the maths. A whole new vocabulary was presented to me and soon the complexities were unfurling like a rose in the early morning sun! However, a run-in with the authorities was inevitable and their unyielding stand, as well as their refusal to listen to reason, made my association with that department an unhappy one. At first I put it down to my tender years, but as time moved on it became apparent that our disagreements were of a personal nature and no amount of scientific proof would sway them from their preconceived conclusions. I still feel that with just a little co-operation great objectives could have been met and both sides could have benefited.

However, the time for recriminations is long past, but as I sit writing this, I feel the anxiety and frustration that I thought were long dead and it makes me unhappy to realise that those years are gone and can never be recaptured. For there is one thing that always wins and that is time!

The culmination of my efforts lay in my garage for nearly two years after the final refusal and slowly, over a period of months, I destroyed that large rocket with a hammer and took the wreck to a dumpsite and deposited it there. It took four trips to the dump to get rid of the entire rocket along with the control panel that had served me so well in firing rockets over the years. I felt at that time that I wanted nothing more to do with the past and therefore was determined to sever all links.

Life goes on and I hope that at least one of my grandchildren will carry the interest into the future, as the exploration of space is now an everyday happening. The future in the third millennium is filled with exciting prospects of exploration to our 'nearby' planets and the hardware for these odysseys is available as well as the scientific equipment to evaluate the findings. The advent of the International Space Station proves that old enemies can put their differences aside and work together for a common goal. Just maybe, I shall live long enough to see these wonders unfold.
LIFT-OFF!

Des Prout-Jones
Linmeyer
May 2001

Chapter 1

The fuse is ignited

Dreams of things far beyond a young boy's capabilities or understanding can be submerged and forgotten or they may, with determination, become reality. Most of us have vivid dreams of what we would aspire to become when we attain adulthood. Airline pilots, deep-sea divers, train drivers and firemen are a few of the lives youngsters dream of pursuing. Sadly, very few of us realise these ideals and we end up in totally different careers. My dream started on a Friday afternoon, a dream that became a stark reality that stayed with me for my entire life. A dream that stayed in my mind every day, and sometimes kept me awake at nights. I say 'dream' because at that time it was something that nobody in the entire country had attempted, and, in my childish imagination, I saw myself doing it.

On that particular Friday I sat in front of the radio listening to a programme about the hero chasing a spaceship, when the germ of a dream entered my consciousness. Listening to the sound effects of the rocket on the radio, I had visions of the flame driving the spaceship at incredible speeds. I had not the faintest idea of how a motor car engine worked, and even less about a rocket motor. But it seemed to be far simpler. Also, the very thought of this means of travel conjured up all sorts of fantasies in my mind. Flames roaring out of the rear of the streaking cigar-shaped vehicle, the stars flashing by, all seemed to gel into one. At the age of 11 I was going to build one of these machines! After all, how difficult could it be to make a round pipe with fire spewing out of the back? Firework skyrockets worked in the same way! Here was the start. I would get hold of one of these things, take it apart, and the answer to my rocket building would then be at hand!

In South Africa, 5 November was celebrated with fireworks to remember Guy Fawkes. This gentleman had attempted to blow up the House of Lords in London in 1605 but was arrested and later executed. His exploit was commemorated for many years throughout the British Commonwealth by lighting bonfires and setting off fireworks. This exuberance was celebrated in memory of the 36 barrels of gunpowder that he had placed in the cellar of the House. Only for a short period prior to the fifth were fireworks available to the public in the centuries that followed. After the cacophony of loud bangs, whistles and whooshes that assailed the ears each 5 November, life returned to normal.

However, for me, it presented the opportunity to obtain the necessary skyrockets for my own little attempt.

On carefully slitting one of them open, I was confronted with a hard, black, short thing! As to what it was, was beyond me. I carefully packed it away and sought help or, at least, enlightenment. This I received from my Grade 5 teacher. She told me that it was gunpowder, tightly compacted, and that I should not 'fiddle' around with such dangerous things. And no, she did not know how it was made or what went into it. Surely some book or person must divulge this information! It was 1947, not the Dark Ages!

By now months had gone by and I had yet to make one tiny rocket. But I was not going to do anything stupid by not knowing what I was doing. Too many people had warned me about the inherent dangers associated with my new driving force and for me to ignore them was unthinkable. My safety could only be assured by one of two things:
- Forget this crazy notion.
- Somehow lay hands on the correct literature and *learn* before I attempted anything in this direction.

Words of wisdom given by sincere folk. My quest now entered into the hallowed, silent halls of the large Johannesburg Public Library where I was shown to the relevant section. Confusion reigned as I turned the pages of book after book. Periodic tables? Specific weight? Pound per mole? What were these things? Help! At last I opened a book and scanned the contents: *gunpowder*! I sat down and carefully wrote down the names of the ingredients. My eyes were sore as well as my nether region as I descended the library steps and made my way home.

Armed with the information that I had sought for so long, my mind was filled with the next steps. I now, somehow, had to obtain the chemicals and what better place to start than at a chemist shop. My enquiries yielded fruit. Yes, they were available, but in small quantities. I continued on my homeward journey far happier than I had been for some time. Plans were forming in my mind, the shape of the rocket, the size, the launch stand, and its angle and supports. Oh yes, now things were moving! In my defence I must say that my schoolwork never suffered because of this interest, but rather improved, because one day I was going to understand all that I had read in the library. I was not going to be satisfied by being dumbfounded by what others had written! I had chosen the path and I was going to walk it!

This was the first of my peregrinations to the library as I sought more information not only in chemistry, but also in aerodynamics. I read books and more books on these subjects until one librarian suggested that I bring my bed and sleep there! This she meant in the kindest possible way. The books that I was interested in were to be found only in the reference section, which meant that I could not take them home. This necessitated my frequent visits to the stillness, and almost reverence, of the hall of the library. I still drop in from time to time to up-date on the latest publications.

Money in those times was not easily come by, so the purchase of the chemicals would be delayed until I could save the necessary amount. However, there were more ways than one to fuel my dream! The short black thing I had saved from the skyrocket! Here was a readymade gunpowder fuel pellet, just waiting to hurl my rocket into who-knows-where. The rocket was ready and so was the launch stand. A few modifications to the rocket to accept the black fuel pellet and let the countdown start!

On reflection, that rocket looked more like an ocean-going yacht without mast or sails, but I had spent hours making it out of a block of wood using a chisel. To me it looked fine as I prepared to saw it down the middle because the hole at the rear was too small to accept the black pellet. A hollow was formed in both halves, leaving an exit in the rear hole. Hot wood glue saw the two sections neatly joined together and once more I was ready to launch. The hollow I had made accepted the pellet perfectly, aligning it with the rear hole. This hole would admit the ignition fuse as well as acting as the exhaust port. Roll on tomorrow!

The following day I had a load of homework to do and by the time I had finished, it was suppertime. I glanced up at the darkening sky and gazed at the moon. This beautiful, round ball held an ever-increasing fascination for me. One day a rocket would take people to its surface, and of this I had no doubt. Even in those bygone days I was convinced in my heart that human beings would eventually break the bonds of Earth and travel in space. I was certain that in my lifetime I would witness a human being stepping onto the cratered surface.

Reading so many books over the months had shed light on how a rocket could operate in a vacuum. Any internal combustion engine depended on oxygen from the air to burn the fuel in its cylinders and produce motive force. Jet aircraft also depended on oxygen from the air but could function at very high altitudes. However, there was a limit to which a jet could ascend before the amount of available oxygen became insufficient to maintain combustion and the jet would suffer a flameout.

This restriction did not apply to a rocket as it carried its own oxygen in addition to its fuel. In a solid fuel rocket one of the nitrates, chlorates or perchlorates provided the oxygen necessary to burn the fuel mix. In a liquid fuel rocket an oxidiser such as liquid oxygen or hydrogen peroxide is carried in a separate tank to the liquid fuel. (These two types of rocket propulsion will be discussed in depth in a later chapter.)

After supper I hurried into the back garden and set the wooden launch stand at an angle, aimed at the moon. Next, I carefully placed my 20 cm rocket on the stand and inserted the fuse. It was a tight fit in the exhaust hole. All those people in the 'know' went to great lengths to explain that skyrockets depended on the bottle in which the stick was placed to capture the flame emitted out of the rear. This, of course, pushed the rocket upwards! (Why did I not think of that?) So, just behind the fuse, I placed a wide-mouthed jam jar. I later found out that the bottle was only a means of holding the rocket in a near-vertical position and had nothing to do with the propulsive force. Everything was now in place and the time was 08:00 time to ignite the fuse. My hand trembled as I brought a burning match towards the fuse jutting out of the rear.

The fuse sputtered into life and I dashed for the safety of the corner of the house. Time seemed to stand still; the muted sounds of the voices of my family filled my ears. The sparks from the fuse disappeared into the rear of the rocket. A brilliant blue flash and a loud bang issued forth! The garden was wreathed in white smoke, obliterating the launch stand and the muted voices now became yells and shouts! I was so terrified that I just stayed rooted where I had taken refuge. Something had gone terribly wrong!

It seemed that all of my family were yelling at me at the same time, although I could hardly see them through the thick white cloud. I was sent to bath and then to bed. I was reeking of gunpowder and it took a long time for the furore to die down and for me to go to sleep.

The next morning I recovered all the bits and pieces and was surprised to see that the fuse was still firmly in the exhaust hole. So this is what had caused the two halves to come apart with such force! I had learnt an important lesson and was very grateful that no injury had resulted. If I wanted to continue with this activity, it would have to take place far away from home where nobody would be at risk. The stuck fuse blocked the exit hole completely, not allowing the rapidly expanding gases to escape. This caused a dramatic rise in pressure inside the glued halves and the little rocket burst at the seam. (The failure point of hot-melt glue had been reached!)

I vowed that my next rocket would not suffer the same fate but be made of stronger material. Also, I was going to change the look of the whole thing. Fins for increased stability would be added, as well as a uniform-looking body. (I had yet to figure this out.) I had seen pictures of the German rocket of World War II, the V-2, and this really appealed to me, as it was a real machine that worked the way I had envisaged. Not the awful destruction it wrought, but the awesome power the take-off and the in-flight image left in one's mind. Here was a true spaceship!

Working with the tools I had available it was impossible to achieve a symmetrical shape, with the result that the next two rockets were not as I would have liked. Believe me, they bore no resemblance to the V-2! However, despite this, they were not too bad. But the acid test came when they were ignited! I had enlarged the rear hole to accommodate the same type of fuse but now, instead of using the inside of a skyrocket, I at last was able to use my own mixture of gunpowder. At this stage my eldest brother, David, told me that since he could run faster than I could, he would ignite the fuse. (Right up until the time that I was stopped from making or launching rockets, he was given the honour at every launch of pressing the firing button on the electronic panel.) While he applied the flame to the fuse I took cover behind a stout boulder. I had moved the tests to a remote *koppie* not too far from my home, so it entailed a walk carrying the necessary paraphernalia

See page 117 bottom for enlarged version

Unhappily, these two slightly bigger versions went the same disastrous road as my first attempt. The only consolation was that there were no verbal repercussions afterwards. The low frequency bangs did not carry far in this desolate area, so nobody was unduly alarmed. Three firings, three failures! A 100% score. The time was now on hand when the motor by itself would be tested and weeks of blisters and hard work would not literally go up in smoke and pieces. The following weeks saw all sorts of weird and wonderful displays of pyrotechnics gracing our lonely hideaway: sparks, coloured lights, whizzes, bangs, you name it. We had them all!

By now I had entered high school and as yet had not managed to get a rocket to lift one centimetre off the ground. I was learning all the time about this beast that kept eluding me, but deep down I knew that success was not far off. Rocket terminology was part of my vocabulary and with it, understanding. I now had all the pieces together of my many failures and I felt that, although the road had been long, the very principles of a rocket motor were known to me. Three years had gone by since I had decided that I would make one of these vehicles and I had read books on the early pioneers in this field. They had suffered the same results as I had experienced and this, more than anything else, encouraged me.

The next year the family went to the Natal South Coast for our annual holiday and it was here, far away from home, that the final realisation impinged on my brain! A rocket motor did not rely on the gases being trapped in a vessel at its base, forcing the vehicle away! How could that possibly be true? The pictures I had seen showed no such thing! The rocket lifted cleanly from the ground. A metal stand supported these rockets, perhaps a few feet above ground level, allowing space for the hot flame to escape from the exhaust nozzle. I saw now that the force was actually generated within the combustion chamber and this was no longer a mystery.

The easiest way to define this action was to imagine four

men in a room. The room was square, with all four walls being equal in length and height. In the middle of one wall was a door. Three men stationed at the centre of the walls without the door exerted an equal amount of pressure against these walls. The fourth man exerted the same amount of pressure against the door. With all four walls being subjected to equal pressures, the room was in a state of equilibrium and all forces being equal in all directions, the room remained at rest. Suddenly the door was opened and that man fell out! Now what happens? The two men pushing at the sides of the room cancel each other out but the man at the opposite side to the door now has no opposing force and the room has a tendency to move in the direction in which he is pushing! Later I found that Sir Isaac Newton had put it very nicely in his Third Law of Motion, stating that for every action there is an equal and opposite reaction. He discovered this around 1679. His Three Laws of Motion helped to make spaceflight a reality 300 years later.

Once I had fully grasped this fundamental, the very core of my dream was opened to me. On our return from holiday my head was full of new ideas, particularly concerning the propellant I had tried so unsuccessfully to use. In the first place, I had no means of compressing the powder mix into a solid pellet and I was certainly not going to try. I had seen too many mishaps with small quantities even to think of pursuing this avenue using larger amounts. The reason for my efforts ending with fizzles and bangs was not the mix of the ingredients being wrong, but the application. I had poured the powder into a small tube, not realising that air was trapped in tiny bubbles in the fuel. When ignited, the flame was not confined to a defined area but had shot into these air spaces, setting off the entire charge at once! Hence the amount of gas was too great for the exhaust outlet to handle and therefore something had to burst, namely the entire model.

The first change to be introduced was the actual combustion chamber. Up to that stage, any suitable small metal tube of whatever diameter had been used. If I thought it was too long, it was cut smaller. If too large a diameter, it was discarded. Getting what I thought was the correct tube proved to be a problem and one that was not easily solved. That is until I had a stroke of sheer genius! I rushed outside and took the pump off my bicycle – it was exactly what I had been searching for. It was seamless steel and it had a threaded front where the handle went in.

Onto the bike and down to the local cycle sales and repair shop! I knew the owner, so felt no embarrassment in approaching him with my request. He eyed me with a strange look when I asked him for any or all the pumps he had no use for. He walked toward the back of the shop where the workshop was, turning around twice to fix me with that funny look. When he reappeared, he had a good half dozen in his hands and held them out to me. I gratefully took them and he asked in a low voice what the **** was I going to do with old pumps? I told him, and I think at that moment my sanity was in serious question. He wanted no compensation for them but added in a friendly tone that he would keep any others that came his way. That gentleman never knew what a great service he had rendered me, because these items would set me on the way to much bigger things.

Now I had the combustion chamber that I wanted and I set to work modifying it. Out with the handle and plunger, fixing two metal washers of the correct size, to cover the hole in the screw cap where the handle had once been inserted. I could now fuel the chamber from the top, closing it with the sealed screw cap. This was more like it! Four fins cut from thin sheet metal were fixed at the rear and a conical wooden nose screwed onto the front cap completed the transformation. This was the biggest rocket I had made and it looked like a rocket! Very well, now make it fly like one! The hard part!

That night I carefully stood it on the small table next to my bed. My dreams, when I eventually went to sleep, were filled with the most incredible rocket launches ever seen.

I had read about two specific powders that, when mixed, gave a safer and more reliable performance than the old black powder. Enquiries revealed the suppliers of these chemicals and a visit to the company nearly gave me a nervous breakdown! Not only did they have them but the manager also told me that for the advancement of rocketry, he would supply them at no charge! For seven years he kept to his word and I am profoundly grateful. This mixture brought a whole new meaning to my efforts and by experimenting with small (2 to 3 g) loads in the open, I achieved the optimum mixture ratio. Now I was prepared to try this propellant in a combustion chamber. I placed the same amount that I had trial-mixed into the chamber and screwed the top cap in place. The fuse gave off an impressive number of sparks and with a loud 'Whoosh' my rocket disappeared over our fence and into the street! I stood thunderstruck, gazing at the spot where the last puffs of smoke issued from the exhaust exit. At last, after more than three years, my bird had found its wings! The back garden was one huge, white cloud and I stood in its midst with tears running down my cheeks. (The tears were not from the smoke!) Where the launch stand had stood, a large grey circle covered the ground. The aftermath of the take-off – residue from the exhaust gas. The rocket had covered a distance of 30 m at an altitude of approximately 10 m. Lift-off! I was airborne. Gone were the memories of the dozens of miserable attempts and the disappointment felt at each failure. I believed at that moment that the genesis of a new breed of rocket was taking place in my life and the smoke that swirled about me seemed to confirm my feelings.

In the three weeks following that launch, I fired the remaining five modified pumps, adding a little more fuel each time. The launch stand was altered to guide the rocket vertically, not at an angle as before. The highest altitude attained was in the order of 40 m and I was highly delighted with the straight ascent of these units. On a vertical trajectory, the rockets came to earth close to the launch, which saved my brother and me unnecessary walks! I must hasten to add that each launch was different, the flash on ignition, the cloud of smoke and, of course, the loud 'whoosh' at take-off. One aspect was common to all and that was the speed at which the rockets left the ground. They fairly leapt into the air!

My supply of bicycle pumps was now constantly being renewed, so different shapes of nose configurations were tried as well as fin shape and placement. I had a reliable rocket motor to test these ideas and my confidence grew by the day. But I never loaded the fuel anywhere near maximum capacity and stuck to a quantity that I knew to be safe and reliable. Rockets streaking into the blue South African sky were a common sight for the next few years. I had invited a few of my friends to witness some of the launches and not one of them failed to be impressed. Somehow I think that one of them passed this onto the press and one evening I was visited by a reporter and photographer. A picture and article appeared in a daily newspaper; the date was Saturday 13 June 1953.

That weekend I did not know whether to be worried or elated. The last thing that I wanted was publicity of any sort, and to avoid this I had carried out my experiments at an isolated area away from any habitation. Safety had always been uppermost in priority since the night that tiny rocket had caused such a commotion. It strongly brought home the reality that these things were not toys and demanded respect. In fact, I had finished a rocket that was nearly a metre long and was waiting to launch it when the press interviewed me. I did not tell them about it as I had arranged to have the firing photographed by a very good friend. He was an amateur photographer of some note and I had asked him to do the honours. His picture of the firing is historic. The first recorded rocket launch in South Africa!

Monday dawned and at half-past eight my telephone rang and with shaking hands, I lifted the handset. A strident voice enquired if that was one Prout-Jones! It was the Inspector of Explosives, from the Government Department of Explosives. Here was the start of a battle that lasted 11 years and one that I was destined to lose.

I was told to present myself, with my father (because I was still a minor), at the inspector's office as soon as possible. I dared not ask questions on an occasion such as this, but hastened to face the music! On the way, my father expounded on the unpleasant consequences that awaited me when I arrived at said office and maybe now I would stop 'messing around'. I could hardly sit still in the waiting room. At last a stern-faced woman said: 'Mr Humphries will see you now.'

Mr Humphries sat behind a large wooden desk with his elbows resting on it. He pushed a newspaper towards me, and I looked at my face staring back at me. There were no pleasantries and he launched into a tirade immediately.

'Have you got a blasting licence?' (This was a prerequisite for miners!)

'No, Sir.'

'Then what do you think you are doing? Not only are you a menace to yourself but also to the public!'

'I am trying to develop a rocket to be used for scientific purposes, Sir.'

At this, the Chief Inspector of Explosives demonstrated exactly why he was in that exalted post! He exploded! I was going to be fined, I was going to jail, I would NEVER dabble with any chemicals in my life again and what was more, my whole idea of making a rocket was STUPID!

After this tirade he slumped back in his chair and eyed me with a piercing glare to make sure that all he had said had imprinted on my peanut brain. Satisfied that I had understood all aspects of my nefarious actions and, more importantly, the awesome consequences that awaited me if I did not heed the warnings, he rose to his feet. As far as he was concerned the matter and meeting were over. I respectfully stood up and was about to head for the door when it struck me that this was no way for years of trying to end. I turned to face the chief inspector and began to tell him of the future projects and vehicles that I had planned, since the propellant I was using was safe. Safe when used under the correct manner and conditions. I stressed the safety precautions that prevailed at every launch. I invited him to attend one of my rocket launches and he declined. But now came the character of the man. If I submitted to him the size of the launch range, the safety precautions to be followed and a complete specification of the rocket, he would consider my written application! I thanked him profusely and swear I saw a masked twinkle in his eye! We left the building with a sense of relief and I knew that I could comply with his conditions.

However, when it came to sitting down and putting pen to paper, it was not that easy. I took time to find the proper words to explain the procedures correctly, leaving no possibility of misinterpretation. The hardest part was ensuring that not one of the safety regulations was omitted as I felt that herein was the crux of the whole exercise. When the documents were complete, I took them to Mr Humphries, ensuring that they landed on his desk and not somewhere else.

I was surprised when I was told that he wanted me in his office. He motioned me to sit down whilst he read my paper. Every now and then, he nodded his head or asked me to answer a question, particularly about the rocket, which was the one waiting to be fired. (The 'big' one.) At last he lifted his head and fixed me with that steady gaze. He would answer my request in writing and I was not to fire without his written consent or refusal. The wait seemed interminable. Was the reply to my request so difficult that it took this long a time? Up until then I had no idea as to how long it took for the governmental wheel to start to turn! After what appeared to me an eternity, the letter arrived! It was in the affirmative but with an important condition. If, by any chance in the future, permission were granted, inspectors from the department would arrive at the launch site to observe and report any deviations from my laid-out safety procedures. Their arrival would be unannounced. In the specifications I had submitted, the proposed date for the firing of the 'big' rocket was appended and now I fully expected them to arrive.

Preparations for the coming shoot were completed and I had set the time for launch at 17:00. To meet at this time would mean we had to be at the ground at least two hours before. This was necessary to set up the launch stand, mix the propellant, fuel the rocket and ready it for firing. (I must add that as time went on and the rockets became larger, this time period to ready every operation became longer. In some instances, a day or two.) On arrival at our special place, my brother and I started the tasks in hand. My friend with the camera, the late Jack Holloway, started taking pictures of the entire proceedings. Later he was to give me these photographs as a memento, little realising the part in history he had played.

Jack was to become our official photographer, but what is more, he was to earn my respect in so many ways. His friendship and genuine interest in me as a person are something that I cherish to this day. He was my confidant, my mentor in electronics, my critic and friend. His amazing pictures will live on, a special epitaph for a special man and I am that much more fortunate in having known him.

Preparations went smoothly with only a few minor hitches to overcome. The first rocket to be launched was to be a 'pump' version to give the few spectators a small taste of things to come. Also, to allow Jack the opportunity to get the 'feel' of the high speed of the lift-off. Every aspect was now completed and with everyone at a safe distance behind a boulder, the match was applied to the fuse. The launch rail was set at 45° so as to present the whole flight as a side-on view.

Because of the use of a fuse to ignite the propellant, no accurate countdown could be given. Only by watching the spluttering sparks reaching ever closer to the rocket could one gain an approximate time of ignition. The motor ignited with a low bang and sent the rocket into a long curving flight, smoke streaming out of the exhaust leaving a white trail against the sky. On landing, the rocket buried itself half-way into the ground, and the part that was exposed was blackened. Silent testimony to the heat of the burning fuel. Prior to the launch I had made it clear to those present that I and I alone would retrieve the fallen rocket and all were to remain where they were until the 'all clear' was given. I waited a full five minutes before going to the grounded casing. With a little persuasion I pulled it free of the clinging soil and walked back to show the small group. All were suitably quiet and I noted that a few were wiping their eyes from the effects of the smoke. Our intrepid photographer was most impressed and from that moment on he never clicked the shutter at less than one thousandth of a second to capture the take-off.

Now came the moment I had been waiting for! The big rocket was something else. I had not attempted anything this large before and also the amount of propellant was more than double that which had hurled the previous one into the air. This rocket was 90 cm long and 3,8 cm in diameter, had four fins at the rear, and was painted white with black lettering and the number '1'. This was the first in my Saxon series! I retained this name for every rocket that I made in the years to come. This motor was fuelled directly through the exhaust nozzle at the rear. A metal disc was welded to the front of the motor tube and a wooden nosecone was attached to this. It was 17:00 and time for lift-off. (Generally, at this time, there was very little wind.)

No official had arrived so I took it that they would not attend this shoot and gave the nod to proceed. I remember the way that my heart hammered in my chest as the fuse sparked closer to the tail. Too late now to have any misgivings! With a loud bang and whistling, the white shape leapt into the air, shrugging off the pull of gravity for a short time. It scribed a neat parabola across the vacant expanse of the darkening sky and plummeted into the ground some 200 m away. We had set a new record for ourselves. Jack's picture of this rocket leaving the launchrail is great! My visitors had many questions to ask and it was dark by the time that I arrived home. I packed the spent casings along with my many other mementoes in a large box, but not before I had labelled them correctly. The photograph of the take-off appeared in a daily newspaper some days later and I still have the press cutting.

See page 115 top for enlarged version

Shortly after this had appeared, I was invited to give a talk to a few members of the South African Interplanetary Society. A great honour for a 17-year-old! I accepted with alacrity although, as one can imagine, with great misgivings. Perhaps this is where I would be thrown to the wolves! This was the forum for the experts, but when the evening arrived and I started to speak, I discovered that the audience were hanging onto every word. It was then that I realised that my subject was not understood! Many in the audience of that esteemed body were here to learn. What an opportunity to address people with the same interest as mine. All nervousness went and I warmed to my subject, explaining the theory behind solid and liquid fuel rocket motors, their advantages and disadvantages. At the end of the talk, questions were asked until it was time for me to leave, as I was about to miss the last bus home! A gentleman stepped forward and offered me a lift to my home as he felt that many more questions wanted to be asked and would I please stay? To say I was pleased would be an understatement. I got home very late but very satisfied that perhaps I had won a few sympathetic ears!

With the successful launching of Saxon 1 I believe that I achieved some measure of credence with the authorities. The results of the launch were forwarded to them and I received an acknowledgement back. No queries, no comment. I took this to be a good sign as I prepared the specs for the next rocket. It was essentially the same as #1 but a few refinements to the fin mounting were added. A series of five of this type of rocket was proposed to prove the stability and reliability of the whole. Permission was granted and all five flights were as expected.

During this period I had embarked upon a career as an architect but was not happy. I therefore changed horses and qualified as a radio and electronics technician instead. It is not hard to understand why I chose this field. To help me develop more sophisticated rockets, of course! My employer was a well-known manufacturer of light and heavy-current products. Here I was able to explore my newly learnt skills. It was here that a whole new vista opened to me in the quest I had long ago set out on. Of greater surprise were the men I met who showed more than a passing interest in what I was doing in the study of rocket flight. As a result of their urging, I was later to form the SARRG.

Before this happened, new designs were on the horizon, breaking away from the old concepts. As I now went deeper into the mathematics of motor design, it was as if an opaque membrane had been peeled away from my eyes. I had been fortunate enough to obtain a copy of the definitive work on rockets up to that time, and believe me when I say that it took a lot of hard work to understand the maths. But the reward was more than worth the effort. Now I could sit down with pen, slide-rule and paper, and really design a combustion chamber with the proper expansion nozzle which would stand more than a fifty-fifty chance of working. If it was a hit-and-miss affair before, now it was pure science.

In the factory next to ours, I met a man, Gordon Wilcox, who was to become an integral part of my life. He was able to do anything with a piece of metal, and he and I put heads together to come up with the launch pad that served us so reliably for a long time. His attention to detail ensured that whatever I asked him turned out exactly as per drawing. He and Jack Holloway were also friends and this saw the nucleus of the group starting to take on a rather formidable amount of talent. In the big picture, no one man is responsible for the final outcome. To each of us diverse gifts are given and if we are able to pool these gifts, only great things will emerge!

Strange how things work out. My friend Jack was also an avid operator of a lathe. He added this to his multifaceted talents and was more than willing to take on the task of producing the first true exhaust nozzle for the next rocket. This one was to be constructed out of aluminium to save weight. The motor tube was 60 cm long with a 20 cm nosecone assembly and four delta-shaped fins to provide guidance. The front of the motor was fitted with a 2 cm thick aluminium bulkhead that was a tight fit in the tube and secured with six machine screws. To complete this rocket, the nozzle was well made out of aluminium bar. It had a 15° divergence outlet and a 30° inlet section. The throat in the nozzle was designed to accommodate a combustion chamber pressure of 500 lbs/in^2. The completed rocket was spray-painted white with the usual black number and lettering. These were to become our chosen official colours when the group was formed.

See page 106 bottom for enlarged version

The launching stand was constructed out of mild steel and would allow a vertical lift-off. This was not the launch tower that Gordon and I had thought about together. That one was designed for much larger vehicles. The present stand would cope with the rockets that we were about to fire. The guide rails were 1,5 m long and had an adjusting device to allow for a 3° angle off the vertical. This adjustment would enable me to take wind strength and downrange factors into account. There was always the possibility that the rocket could land close to its point of departure therefore a degree or two pointing downrange was an additional safety precaution. If launched at 90° the returning rocket stood a good chance of landing on us!

Enter the remote launching system. At last the match and fuse age was gone! A full roll of twin electric cable measured almost 100 m and I chose this distance away from the launch stand to ignite the motor. I am sure that this pleased the inspector, as he saw that I was serious about what I was trying to achieve. The added safety built into the project by being that far away when launching was welcomed by all of us. It also provided all concerned with the exact time the rocket would spring to life and not the nail-biting wait for the fuse to burn!

The control panel was equipped with switches and indicator lights to signal which circuits were activated. The most important piece of the entire system was the safety key that I carried in my pocket. Without this key, the firing circuit could not be armed so it precluded any accidental start of the rocket while the launch crew were still busy at the stand. Only when the area was declared clear, did the final countdown begin and at T ('T' is short for take-off) minus 10 seconds would I lean over the firing officer's shoulder and insert the key into the panel. This would now enable two separate circuits to be turned on: the low-voltage heater to the igniter, and finally to the high-voltage firing circuit. This circuit delivered 90 V to the igniter and combustion was immediately initiated. As the rocket left the launch, all switches were returned to the 'Off' position.

Because of the number of essential pieces of equipment, it was not practical to carry them to the *koppie* where we had fired so many rockets over the past years. We had to find a large tract of open space devoid of any habitation. It had to be fairly close to Johannesburg where we all lived. The offer from the manager of an airfield on the outskirts of town came out of the blue! I grabbed the opportunity to visit him and explained in detail what the aim of our research was all about. He listened to me and agreed that we could launch from the very centre of the property, giving us about a mile of nothing in any direction. One condition! The rocket had to be back on the ground by 08:00 before any aircraft took to the air!

A date was set for the launch of our aluminium rocket and all the required permission was obtained. The manager of the airfield was there to meet us and stayed to witness the proceedings. I was aware that the rocket on the launch stand looked anything but dangerous but was quick to realise that if it worked as predicted, the whole picture would change. The cable was connected to the panel and the igniter was inserted into the nozzle of a rocket that was fully fuelled for the first time. The launchrails were inclined to 89° facing due west, and the cable was finally connected to the twin igniter wires hanging from the nozzle. The rocket crew did a final inspection and headed for the control point. We had come a long way since the night I had nearly frightened the family out of their wits!

Time was on our side as the flight co-ordinator called out the remaining seconds and sequences:

'T minus 15.'
'Master?' (Flight)
'On.' (Firing officer)
'Power?'
'On.'
'Heater?'
'On.'
'Rocket clear!'

I inserted the safety key into the panel and sat back.

'5, 4, 3, 2, 1, Ignition!' The firing officer brought his thumb down hard on the firing button.

The ignition was the loudest any of us had heard and I wondered what those who had never seen a firing must have thought. Has the whole damn thing blown up? Smoke and fire! However, it was the sweetest sound I had heard. My eyes were fixed on the slim, white shape streaking up and up. It reached its zenith and slowly the nose pointed earthwards. It seemed a lazy turn as if reluctant to return to its lift-off point. The two men tracking the rocket with a type of inverted protractor wrote down their readings. (They were stationed at a distance of 100 m from opposite sides of the launch. From here, two angles were measured at the apex of the flight. By knowing the distance from the launch and the tangents of the angles measured, the vertical height could be calculated.)

The flight's voice sounded strange as he called the end of flight procedure: 'Heater off, power off, master off'. His job was completed and he could not hide the nervousness in his voice as he glanced at the panel to make sure all his instructions had been carried out. This was the baptism that would be faced by this team each time we launched. We had come together through a common interest, had held simulated firings and had performed well as a cohesive unit. The tracking men handed me their pieces of paper with the angles at apex they had recorded. I did the trigonometry and gave the altitude as 1 000 ft taking it to the nearest 100 ft. The calculation was close to 1 100 ft but I chose the former. This was the first of our really successful rockets in all aspects.

The cable to the panel was disconnected and all of us walked to where the rocket was sticking out of the ground. Nobody spoke as it was retrieved and found it to be intact! The only damage was a slightly dented nosecone. A bit of panel beating and this 'bird' could fly again. The newspaper the next day was most kind to us. It seemed that our efforts were gaining interest. The editor of a daily newspaper spoke to me asking to be invited to all future firings! The next shoot was to be an attempt to attain an altitude of a mile and this was a challenge to me personally. It had not been attempted by anyone in South Africa and if I succeeded, history would be made. To my knowledge at that time we were the only people to seriously attempt scientific rocket launchings. However, I was not going to pass up the chance to be first to reach that goal.

Chapter 2

Reaching for the elusive mile

One mile – 5 280 ft (1,6 km), not a long distance if you were travelling by motor car or a bicycle, or even walking. An adult human being walking at a normal pace would cover the distance in eight minutes. Yet to make a rocket with our limited resources reach a mile, a single mile, was a target that I had in mind ever since the 'big' rocket swept across the late afternoon sky. The sight of the last one climbing effortlessly into the early morning sky seemed to say: 'You can do it.' There were a few reasons I could think of that prevented me from doing so at that time. Foremost was the propellant that we were using, although it had proved reliable in so many launches, the power output per pound of fuel was low. Therefore to go higher would require a substantial increase in fuel, which in turn meant a far bigger, heavier rocket. It was like a dog trying to catch its own tail! Round and round. The solution was staring me in the face and as yet I did not see it.

To achieve good results in the design of a rocket the mass ratio, that is, the weight of the fuel compared with the weight of the complete rocket should be in the order of 80% to 20%. That is, only 20% should be structure. This parameter is not easy to achieve and the search for more powerful propellants is the logical step. But the more powerful propellants require stronger and heavier materials to ensure safe operation. So, on with the dance! There had to be a way where all sides of the equations could be satisfied and perhaps a revolutionary design would yield the solution.

This came to pass. It was called the Bumper/ Wac Project. A V-2 rocket was fitted with the much-smaller Wac Corporal rocket on top of the nose. When the larger rocket was fired, it carried the smaller one up to the point where the propellant of the V-2 was exhausted. At this moment the speed was at its greatest, the Wac Corporal was started, adding its speed to that of the V-2. This final speed was almost double that of the V-2. The result was that the altitude was more than doubled! Enter the multi-stage rocket. It was 1949. The altitude achieved by this combination was 250 mls compared with a V-2 rocket's altitude – 115 mls! This altitude record remained for some years. The design had met the expectations of many designers and without realising it, the door to space flight was being opened.

The mathematics suggested that if the mass ratios of the different stages were maintained as close as possible, the actual doubling of speeds was achievable. This concept presented me with the very system to achieve an altitude more than five times that of the last rocket we launched. So, to work! The calculations were long and time consuming but at last a set of working drawings were made. These were given to Jack, and slowly the embryo was formed. It was noticed that severe erosion had taken place in the last nozzle, scouring the aluminium badly. It was necessary to use mild steel for this new rocket and I was amazed to see more than three quarters of the steel bar turned into swathe as the nozzles took shape. The scouring we observed was caused by solid particles in the exhaust flame passing through the nozzle at supersonic speed.

Great what correct design can yield. Essentially the bottom (or first stage) was based on the previous rocket, but somewhat larger, and the second stage was slightly smaller. The nozzle of this stage was a telescopic fit into the top of the first stage. This rocket was designed to carry a small charge of flash powder in the upper stage where a mercury switch joined to a battery would ignite it. This would take place as the rocket turned over on its downward path. A small white cloud would then suddenly appear high in the blue sky, marking the height that the rocket had reached. This made it easier to get a fix for the men tracking it for at that height the rocket would be invisible to the naked eye.

To ignite the second-stage motor a rather simple method was devised. The height at the first-stage motor's cessation was known and an appropriate length of piano wire was attached to the launch and the other end of the wire was attached to a detachable plug on the side of the upper stage. When this plug was detached, it enabled a switch to close, discharging an electrolytic capacitor's 360 V into the second-stage igniter. This capacitor would be charged on the ground prior to lift-off. The charging batteries were housed in a special box and were activated from the main control panel. This high voltage was routed to the base of the rocket along a separate cable that would detach at lift-off.

The rocket stood 1,6 m tall and looked very professional, but the new launch that Gordon and I had designed made the set-up complete. This was the launching tower that would be used for a long time to come. The base was four pieces of 10 cm angle iron and a heavy cross-member in the centre supported a 12 mm thick iron plate with a 15 cm hole in its centre. Fixed under this hole was the exhaust duct that would allow the rocket exhaust to blow out sideways. The rocket was positioned directly over this hole with the exhaust nozzle just at its entrance. The exhaust duct was to keep the very hot flame from scorching the tail and fins of the rocket at the moment of ignition. The heavy structure had four adjustable metal legs to align the flight path according to ground wind speeds but mainly to aim the rocket downrange. Two 4 m x 5 cm metal poles acted as guides for the rocket and were attached to the pad a little to the side of the exhaust duct.

We arrived at the aerodrome at 04:30 on a chilly Sunday morning and had a hot mug of coffee whilst waiting for the rest of the crew to arrive. We did not have a long wait, and had time to mutter a few derogatory remarks about the time and weather. Then we began the erection and alignment of the launch tower. That took time! Remember we had to have the rocket back on the ground before 08:00. The tower was assembled and aligned due west to afford the longest distance to a boundary. The sun was making an appearance behind us and offered no warmth as the fuelling operation began. This was my job and I was doubly cautious, carefully measuring the propellants for each stage. The period to allow all air to escape from the motors was strictly timed before the two stages were coupled together.

With the rocket now fuelled it was placed on the thick iron pad so I could install the flash powder device in the

See page 111 for enlarged version

forward section of the upper stage. The manner in which the rocket is placed is interesting. Painted on each of the four fins are the letters N, E, S and W. These letters are the four points of the compass and the rocket is positioned on the launch with the fin marked W pointing to the west. The other three fins will automatically be faced in the right direction. This is of importance because when the rocket leaves the tower it must not start to spin but rather fly absolutely straight. By looking at the post-flight pictures any spin may be detected and the fault corrected. Herein was one of our long-term objectives, for it was the intention to mount a camera in a future vehicle. If the camera lens were fixed so that it looked, say, out of the north side of the rocket, it would not be pointing in the right direction if the rocket began to spin. We could end up taking a photograph of the empty sky!

See page 117 top for enlarged version

Up to now I had not had an unstable flight and today, especially today, I wanted a good flight because the press was arriving and also because of the importance of the first two-stage rocket in South Africa. The press photographer was a quiet man and said that if he, by any chance, got in the way I was to kick his butt! I jokingly replied that it would be my pleasure! The cable to the control panel was laid out and connected. Another cable lay neatly next to the main cable and there appeared to be a mess up at the tower, where all these wires had to be connected to the exact circuits in the rocket. The igniter in the first stage was connected last and with that, the mess was now a neat-looking bundle of different coloured wires.

The control panel was connected and it seemed very complicated with the additional high-voltage box standing next to it. There was time for the firing officer and myself (I had been given the title of launch conductor) to do a final inspection at the rocket, making sure that it was as correct as possible.

We had set ourselves a task and had completed it well before the time deadline. By this time the sun was well above the horizon and we had a last mug of coffee from the thermos.

Now it was time for things more serious! A familiar voice called out:

'T minus two minutes and counting.' I knew that the capacitor in the rocket would take at least 30 s to charge to maximum but I told the firing officer to switch on the charging circuit just to make certain. The needle of the milliammeter swung up, indicating that the capacitor was drawing current and charging. As it retained more and more voltage, the current draw became less and this was evident as the needle began to fall. I nodded to the flight and his voice intoned the countdown and sequences.

The final seconds were always a tense time for me and it was something that stayed with me at each firing of all the future rockets.

'5, 4, 3, 2, 1. Ignition!' The moment of truth! The rocket was completely enveloped in a cloud of dense white smoke as the motor started with a loud bang. It hurtled upward and almost immediately a second bang heralded the upper-stage ignition. The rocket appeared to leap as if it had been scorched and then seemed to stand still, growing smaller until it was lost to human sight. I kept my eyes trained on the spot where it had disappeared and I counted the elapsed seconds to myself. Anything less than 14 seconds would mean that the rocket had not reached the design altitude: 12 and 13 and 14 and 15 and ... a new white cloud burst in the heavens! My nerves were at breaking point as I waited for the tracking crew to hand me their measured angles. A thin vapour trail headed towards the ground marked the rocket's decent.

It landed about 1 800 ft from the tower and we had to look for the first stage because all of us had our eyes on the upper stage. When both had been recovered, they were like night and day! The first stage was burnt black from the exhaust from the upper stage whereas that stage still wore its white coat of paint. The post-launch procedures had been carried out and the tracking crew handed me the results.

I somehow knew before doing the calculation that the

See pages 108, 109, 110 for enlarged versions

rocket had cleared the magical mile. It had!

We were elated and I believe that everyone had his hand shaken at least three or four times.

Our joy knew no bounds. This time our rocket design had really paid off and the reporter broke into our reverie asking quietly: 'When is the next launch and how high?'

I turned to him and hugged him in sheer happiness! 'Soon and very high!' I replied.

His pictures appeared on the front page of the paper the following day with a complete write-up under the heading: 'Rocket soars over a mile'.

See page 118 for enlarged version

I was rapidly starting to understand rocketry, and the marvel of it all was that there seemed to be no limit. It encompassed so many different fields of science such as thermodynamics, metallurgy, electronics, mathematics and aerodynamics. More were to be added to this impressive list as the rockets became larger and more complex. After the mile-high shoot, I had the thought in mind to design a large single-stage vehicle with a powerful motor. Far more powerful than I had ever imagined! I had the knowledge and the facilities plus an enthusiastic group behind me, so I decided to do it.

I learn an important lesson!

It seems to be the very nature of rockets to capitalise on human error or oversight. For this reason, everything was checked and re-checked to avoid any mishaps. No matter how thorough you may be, 'Mr Murphy' is always there, peering over your shoulder! One tiny slip, and in he steps. If it happens, it is never pleasant! In all honesty, safety was my prime concern and only by being very careful in every aspect, from drawing-board to launch, could we have a reasonable chance of avoiding accidents. So far, we had been extremely fortunate. The frightening, thunderous bangs of my early efforts had never been forgotten and heaven knows I never wanted to repeat them.

The latest design was a monster! It stood 2,44 m tall,

was 0,76 m across the fins and weighed in at over 60 kg at launch! Here was a vehicle! To facilitate ease of sight against a blue sky, I deemed it necessary to paint it yellow with black barber-pole stripes. The exhaust nozzle was carefully designed to allow for the greater operating pressure, as was the forward bulkhead. These important components were fitted to the motor by means of stout bolts and firmly tightened. Likewise, the nosecone and forward instrument section received the same fastidious attention.

The necessary red tape was cleared and the date for launch was set. With such an untried rocket it was my decision not to invite the press. I took this decision after very careful consideration. We could not afford any adverse publicity, as that would certainly jeopardise our future with the inspector. Better to play it safe. If any officials arrived on the day we were to launch, let them be first-hand witnesses.

If something went wrong, I would rather let them see for themselves our safety precautions and, more important, the rocket itself. Although I had submitted a full, detailed specification of the rocket to the authorities, it was my belief that they should see it for themselves. One could draw funny speculations from a newspaper article! I put my decision to the group's vote and it was settled. No newspaper!

I did not have any misgivings, as they had been invited to attend every shoot since I had been hauled over the coals all those years ago. With this in mind, I went ahead with the impending launch. I was now over 21 years of age and could speak in defence of any decision I made. The Sunday arrived, cold and dark at 04:30. This occasion, the airfield manager did not attend, I think he trusted us and his bed was warmer! The usual procedures started to take place as the various men turned up, all wearing warm jackets and breathing vapour in the chill morning air. I was politely asked if I had gone insane, getting them out of bed at this hour! I often wondered myself!

The fuelling of the rocket obviously took longer because of the greater quantity required. I must have cut a dashing figure standing on the back of my van to reach the back of the rocket, which was pointed nose-down! I waited 30 min before we moved it onto the launch tower to ensure the propellant had settled and that no air bubbles were present. The sun began to rise and shed ghostly shadows as we moved about, completing the last preparations at the tower and at the control panel. Jack held his camera under his jersey in an attempt to stop the lens from misting and muttered something about the lack of light. He could not take any pictures yet and told me to hold the launch if the rocket was cleared before time to ensure adequate light conditions for high-speed photography. The tall pointed shape gradually became clearly visible and cast a long shadow, pointing west.

We were now waiting for the ambient light to improve and all auxiliary lighting had been turned off. The two tracking men were in position 300 m on either side of the tower. All was ready and at 07:45 the flight called the remaining time to launch.

'T minus two minutes and counting.'

Still time to call a halt. The range was clear, the sky was blue, and the air was cold. I looked at the control panel as each indicator lamp lit up on time. I gave the firing officer a nod and inserted the safety key into its socket.

'Rocket clear.' The tall Saxon Alpha stood bright and gleaming in its yellow suit.

'3, 2, 1, Ignition!'

The firing officer pushed the appropriate switch. Now things happened fast! The red indicator lamp flashed bright as electric current flowed along the 100 m of cable to the pyrotechnic igniter in the tail. A white cloud of smoke shot out of the base of the tower. It was immediately followed by a tremendous low-frequency detonation! We had grown accustomed to the rocket's loud 'Whoosh' on ignition but this was something else!

See page 104 for enlarged version

A black piece ripped upward out of the great blanket of smoke that engulfed the tower. The two long guiderails that supported the rocket could not be seen in the smoke and at that moment I knew that the rocket had suffered a serious malfunction! The object that I had seen leave the ground was now on its way to Mother Earth and, because of the distance, struck the ground quietly. During all of the simulated firings, the crew was instructed to remain at their posts until the 'All clear' was given.

Nobody had moved since the dreadful detonation and a terrible hush seemed to prevail. Even the flight was speechless as the firing officer switched off the control panel in the correct sequence. The trackers joined us at the control point, not speaking.

We had confidently planned and built this rocket and something had been overlooked. I walked slowly to the tower. It was no more. The guiderails lay on the ground some 10 m from the pad. The 6 mm steel plate that supported the tail was bent and misshapen, silent evidence to the awesome power of the motor! All the pieces were collected and the post-mortem revealed what had happened.

On ignition, the starting transient had proved too much for the forward bulkhead and the aluminium pipe of the motor had torn loose from the bolts that secured the two. This resulted in the entire forward section plus nosecone being dislodged by the high-pressure gas from the burning fuel. It was this section I had seen burst out of the dense smoke cloud. I had lost a rocket, for nothing was worth trying to salvage. When the nose housing shot skyward, the motor section and fins were propelled, with equal force, downwards. This was the cause of the damage to the tower and the pad.

Our rocket was lost and I had learnt a very hard lesson. As could be expected, the inspector was totally sour about the incident and belaboured me when I presented him with the report. My only redeeming feature was the range safety that was in force. He went on endlessly: 'What if, what if, what if ... ?'

The truth was, and he knew it, that we were pressing in a totally new technology and malfunctions were part of the deal. Nobody in the whole world had a 100% track record in this field. I had presented him with results of close to 100 static firings and almost 500 flight firings. No results were ever fiddled but presented as they turned out. (A static firing is when the motor is bolted down and cannot move. It is carried out to test the design of all main motor components

See page 115 top and page 119 for enlarged versions

whilst firing.) The static run of the big motor had gone smoothly and when disassembled after the test, no suspected or visible deterioration was observed.

After this long session I felt drained and looked forward to a good night's sleep. Yesterday's mishap was passed and went into the history book as one big 'boom'. That night I made up my mind to intensify my efforts in the hardware design even more than before!

Two months went by and in that time I approached a manufacturer of solid drawn steel pipes. I had seen that with the bigger rockets on the drawing-board, aluminium would not stand up to the stress that would be imposed. The steel pipe offered a solution to the high pressures that the new designs demanded. It was my intention to increase the combustion chamber pressure to 68 atm. The steel pipe had a burst pressure of 102 atm. This would give me a safety factor of 1,5. The pipes came in different diameters, from 2,54 cm up to 15 cm. The manager of this company said that he would donate whatever I required free and gratis! As I shook his hand and expressed my appreciation, he put his other hand on my shoulder and wished me luck and good shooting. I had made another friend, very excited to play a part in my research. I would eventually gain a total of five sponsors who helped me reach 'great heights'. My income could never cover the cost of the rockets yet to come and these five sponsors had gladly stepped into the breach. They did not want any publicity either!

The new rocket was almost identical to the one that blew its 'top'. The calculations showed that it was capable of attaining an altitude of a little more than 1 250 m and I was very happy to be able to send a vehicle of that size to that height! If successful, it would certainly indicate that the design parameters were sound and the way to the stars was not just a pipe dream.

The completed rocket stood 2,48 m tall and because of the slightly longer length had a different fin configuration from the last one. (I will explain this later.) It also wore the yellow and black coloration to assist the visual tracking. Somehow this fellow looked as if it was just waiting to break free and soar into the sky! The launch tower had been repaired and looked as if it had never been blown apart. The base on which the rocket stood had also been strengthened by adding an additional cross-member to support the guiderails. Saxon Alpha 18 was constructed from steel and aluminium. A far cry from all previous rockets that we had built, they had been made mainly from the lighter aluminium material.

Fuelling such a large motor required special precautions. As in all our previous shoots, the metal casing of the rocket was attached by an electrical cable to a metal spike driven into the ground. This cable effectively prevented any static electricity from developing during the fuelling operation. Only when the last man left the readied rocket was the cable disconnected. This precluded any electrical spark from causing an accidental ignition, therefore it was the last item on the check-off list to be carried out at the pad.

Another safety feature was the use of gas masks for the rocket crew during fuelling. David and I had developed eye irritation from the large amount of propellant that we loaded into the last vehicle. In fact, I was barely able to witness the lift-off because my eyes were watering so badly. Another lesson! The 'schedule of events', the list of all operations required for the launching of each rocket, stipulated that only two persons were allowed at the fuelling. Myself and one other. It was the event that carried the most risk as the rockets increased in size.

See page 95 for enlarged version

The launch date for Saxon Alpha 18 dawned, and driving to the airfield I was aware of the major advances that had been made during the years. What had started as a childlike dream was now total reality. So real that it felt as if my life had been steered by a higher force along this uncharted path. By now I was a married man and my wife, Pat, was part of this exciting path that was heading ever upwards. My thoughts turned to the 15 active people who had given of themselves to assist in making my dream become reality. These were the nucleus of the SARRG.

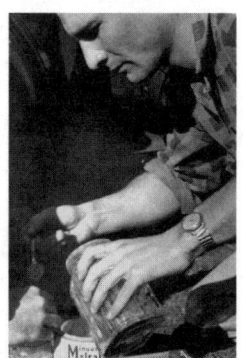

See page 121 for enlarged version

The SARRG had been born out of the fiery exhausts of rockets streaking higher and higher. The ultimate goal was to place a satellite in orbit. A satellite that would circle the globe, transmitting 'South Africa' 'South Africa' to the world! An impossible goal to attain? I did not think so and neither did my colleagues! What is life worth if one does not set a target to achieve and gives up at the first signs of adversity? To persevere and strive to attain that target or goal or whatever is the very life force that we humans have been gifted with! To ignore it is to accept defeat and apathy.

I arrived at the airfield to be met by the manager. He was not going to miss this launch. Other cars began pulling into the launch area and the setting-up started. The various crews gathered together: tracking, range safety, rocket, tower and launch control. The press were busy setting up their cameras as fuelling began. Remembering the eye irritation, I was thankful for the gas mask even if it felt strange. The range safety officer was busy scanning the entire area with high-power binoculars, making sure that no one was wandering around in the restricted parts. All the personnel knew exactly what was expected of them and their tasks were completed efficiently and thoroughly. Everything went off smoothly, and in what seemed record time we were at the control point and the final count had begun.

See page 98 for enlarged version

I looked at my clipboard one last time to make sure every item was ticked off as having been completed. They were. The flight officer's voice broke the early morning silence, not even a bird-song was heard. I inserted the safety plug into the panel and watched the green light appear on time.

'Rocket clear.'

'Ignition.'

The motor ignited with a fabulous, ear-splitting roar! The rocket speared out of the tower at great speed, rapidly ascending into the clear, blue sky. The smoke that erupted from the nozzle appeared as a thick white finger supporting the large yellow shape atop it, seeming to push it ever upwards! (The picture of the lift-off showed the smoke track still touching the ground when the rocket was 100 m off the ground.) As the rocket climbed away, it gently and slowly undulated from side to side, indicating that the dynamic stability was a little overdamped. This condition was caused by my being over-cautious in the design of the fins. They

See page 96 for enlarged version

were a bit too large. Other than that, the flight proceeded as planned.

The rocket started on its downward journey, a thin streamer of residual smoke issuing from the tail. It struck the ground with tremendous force, snapping in two. From the control point we could see the two halves on the ground, wisps of smoke gently curling into the air. It was a good flight and I waited as the trackers brought me their results. It was 1 250 m! We had brought an original design back from destruction and made it perform as predicted!

This rocket had proved that bigger and better performance vehicles were now in our grasp. The motor had started and burned perfectly, producing a thrust in excess of 2 225 N. The way to producing a one ton thrust unit was feasible. When I handed the flight results to the inspector, he asked only: 'How big is the next one going to be?'

Chapter 3

Of mice and rockets

One would have thought that the manager of the airfield would have been suitably impressed after that impeccable flight. Wrong. He called me aside and quietly but firmly said that we had to find another ground from which to fire any future rocket! He had seen this one perform and trembled to think of what could be forthcoming! Furthermore, he had obligations to the owners and he now had visions of seeing at least a runway magically disappear in a cloud of white smoke! I accepted his misgivings and expressed my thanks for his assistance in allowing us the temporary use of the field. So, no more fun and frolic at the airfield!

The old problem had returned, and its solution had to satisfy many criteria, especially the inspector! We in the group individually attempted to locate a suitable site and it was Gordon who caught the big fish – a farm to the south of Johannesburg and owned by a friend of his. I studied a topographical map of the farm with Gordon and the owner and was elated to find that it was ideal. I visited the inspector and showed him our proposal. He hummed a bit and then gave his consent. Mind you, he strongly emphasised the existing conditions were to be strictly enforced! He expounded upon the fact that even though this launch site was on a private farm it did not absolve me from the law as laid down by his department. I had to abide by it, chapter and verse! Or else!

I was busy with the design of a 2,5 m three-stage vehicle when I received a request from a group of biology students at university. They wanted to study the effects of high g forces on a pregnant mouse! Here was a strange one. I personally had not given any thought to such a thing but the aspect intrigued me. To recover a living creature from a rocket flight successfully! When next they contacted me, I agreed, on the understanding that they supplied the passenger. I submitted the usual to the authorities and got the go-ahead. No problem with the mouse.

Apart from the actual rocket design, the compartment for the mouse and its recovery proved to be not so easy. I was given the length and weight of the rodent to assist in the exact placement of the capsule on board the vehicle. It would obviously have to be fixed in the forward section behind the nosecone housing, away from the heat generated by the motor. In addition, a parachute plus its release mechanism had to find a home in the near vicinity. Available space and weight are ever-present gremlins in rocket design and are not always easily overcome.

During the next few days we discussed the precise objectives the students were aiming to achieve. The prime feedback they were after was to learn more about the effect of positive g forces and then weightlessness on a mouse. In addition, this mouse was pregnant so would this damage the unborn? To clarify the meaning of g forces, here on earth we are all subjected to a force of 1 g, in other words, the force of gravity. This is the force that attracts all matter to the centre of our planet. At sea level a mass of 1 kg is attracted to the Earth with a force of 9,8 N but diminishes as height above sea level increases. It has been accepted that at 200 km altitude the region of space has been reached and the condition of weightlessness is experienced.

Further out into space the effects of the Earth's atmosphere are so negligible that artificial man-made satellites are able to orbit the earth without any means of propulsion. They are operating in a vacuum and therefore very little friction is encountered to slow them down.

If we take a ball on a string and swing it around, it will move outwards in a circle. The faster we swing the ball, the greater the outward centrifugal force and the more horizontal the string becomes. Now consider a satellite travelling around the Earth at an altitude of 200 km and a speed of 28 000 km/h. At this speed the centrifugal force will just balance the Earth's gravitational force and it will be 'weightless'.

It is interesting to note that the gravitational attraction of any celestial body never reaches zero before the influence of another body takes over and attraction to it becomes dominant. But back to our satellite, if we impart sufficient speed (velocity) to it at various altitudes we see that the lower the altitude, the higher the speed required to maintain orbit and the higher the altitude, the slower the speed. Conditions in earth orbit are strange in two ways: there is no up or down; and everything becomes weightless. Back in the early sixties when space exploration was in its infancy these facts were as yet not fully understood although they had been known since Newton. These young students were looking ahead in search of some of the answers – hence the flight of the mouse. Mice had been launched in an Aerobee rocket in the USA but the final results were not published at the time. First reports seemed to indicate that the experiment was a success. So this spurred the students on!

To complete their study, the rocket would achieve weightlessness at the time when it arced over prior to its descent. At the end of this period the parachute would open, placing more stress (gs) on the occupants. As an example, if you were subjected to a force of 3 g and your normal weight was 75 kg the resultant weight would be 75 x 3 = 225 kg. In the case of the mouse it would be subjected to 5 gs for a period of 1,1 s. The weight of the mouse was 25 g. So by calculation it would weigh 25 x 5 x 1,1 s = 137,5 g for 1,1 s.

I knew that jet pilots were strapped in special seats to allay the effects of heavy g forces when executing tight turns so as to prevent black-outs as the blood was drained away from the brain. The optimum position was supine. This now showed me the way to construct the capsule and to assist the mouse when the going got tough! A thin aluminium cylinder was made, lined with soft foam rubber, sealed at one end and equipped with a clear plastic window at the other. The capsule was mounted in a dummy nose housing and it contained a thin glass strip inside to simulate the mouse. The glass inside the capsule was still in one piece after being dropped six times from a height of 17 to 20 m. These drop-tests gave me the confidence in the design. The mouse was somewhat rounder than a flat glass strip and would therefore be a better fit in the foam-lined interior. The capsule would be mounted with the mouse in a horizontal position just aft of the nosecone. So much for aerospace medicine just proving how much we did not know.

The rocket to be used was to be powered by one of our reliable 1 780 N thrust motors. We did not want a big bang on the launch pad! I received a telephone call from a friend who worked at one of the daily newspapers that attended most of our shoots. He asked what was in the offing and I replied that no big job was planned only a low altitude shot with a mouse as passenger. He asked to be invited to our next big one and I said: 'By all means.'

I should have kept my mouth closed! The headlines that night read 'Mouse to be launched in Transvaal rocket'. I was stunned! Nobody could put that amount of importance on it while around the world wars were being fought and people were dying! I was wrong. By the time the media had finished with me, every newspaper, newsreel and radio station had got onto my little red wagon! Boy, did they get their pound of flesh!

Work on the rocket proceeded well. There were numerous changes to the outward appearance as could be expected and when completed, it was beautifully sprayed a bright yellow. I sat looking at it for a long time, not failing to admire this 1,83 m piece of high-powered vehicle. I came to the conclusion that the safest way to eject the 'chute was our old and trusted method, using a mercury switch and battery to ignite a small charge of flash powder. Just enough to jettison the capsule and 'chute. It had worked very reliably in the past and there was no reason that it should not perform now. To be honest, the amount of interviews and publicity I was receiving scared the living blue-blazes out of me! All I wanted to do was design and fire rockets! Not be hung out to dry? (When the two-stage rocket exceeded the mile, I received a page from the journal of the Reaction Research Society in the USA showing the rocket in the tower with yours truly next to it. I still have that page to this day.)

If I had an attack of nerves before, it was nothing compared to what I felt on the morning of the launch! It seemed as if everybody and his uncle were there. The range safety officer had his hands full and I wished for the very first time that the authorities would arrive. They did not. It was a February morning and by 07:30 I could feel the sweat running down my back as I laboured to complete my work. In the end I stated flatly that if unwanted people insisted on poking their noses in where they were not wanted, I would call off the shoot! This statement brought a semblance of order.

In retrospect, I could understand the jostling and manoeuvring to get the best pictures or footage on film. However, I do not think that those present had the slightest idea of what was involved. Not only that, but had I not called a halt to the proceedings, the breach of safety would have placed me in an invidious position with the authorities. It was like having an unruly mob cavorting around!

At last I was able to get the mouse into the capsule and safely aboard the rocket amid a flurry of flashes from the press. The members of the crew not engaged in any duties were quickly roped in to keep the media well away as the fuelling operation began. This took place at a distance of 100 m plus from the closest person because of the inherent danger that went with this operation.

I knew from experience that this would take the best part of half to three quarters of an hour to complete. It could not be rushed and we had to load the exact amount of propellant into the motor barrel. While this was being completed, the cable to the tower had been installed but not connected. The tracking crew had taken their positions and were ready. At last the rocket was moved onto the pad. I had already briefed the cameramen and reporters on the countdown sequence, so they knew when to expect the actual firing. A last check at the rocket, the mouse and the parachute release circuitry. I signed the checklist on my clipboard and took my station behind the firing officer at the control panel. All connections to the panel by this time had been completed and everything was as ready as it would ever be. The flight's voice broke the silence.

'T minus 15 seconds and counting.'

The remaining seconds were intoned in the usual way. No holds.

'... 3, 2, 1. Ignition!'

The roar of the motor start was heralded by a burst of flame and smoke from the tail that immediately shrouded the tower from view. The rocket cleared the huge smoke cloud and headed straight up into the blue. A sound of applause came from the spectators but I felt anything but happy! I had seen an object eject from the side of the ascending vehicle and straight away realised what it was! The 'chute had deployed early and was ripped to shreds as it unfurled into the high-speed slipstream. The rocket carried on upwards, hardly affected by what had happened. It reached its zenith and lazily turned, nose down, and began the long fall back to the ground.

It struck with a dull thud, smoke drifting skywards. I reached the fallen phoenix and hastily withdrew the capsule and emptied the contents into my hand. The white mouse moved about in the palm of my cupped hand for a few seconds and Jack was able to capture this on movie film. Suddenly it stopped, then quietly expired. I felt like weeping. As far as I could tell there were no bones broken or organs ruptured. No blood. But the little creature was dead.

'Des, it reached 1 220 metres,' It was the voice of one of the trackers. 'I did the calc. But you can check it.'

'No need to, thanks,' I replied.

Now the boys with the cameras and notebooks were onto me like bees around the proverbial honeypot! Questions flew thick and fast, and I did my best to answer them honestly. Inside my head, the answer to the premature 'chute release came to me. The mercury in the switch had moved forward, owing to the deceleration, when the motor had ceased thrusting. The mercury had made contact with the two metal contacts of the switch, causing the flash powder to ignite. This knowledge was of little consolation now.

The following morning the kindest headline read: 'Roger, the victim of science', with full-page photographs. Other papers were not so vague and where the name 'Roger' came from, I had no notion! I suppose it added that bit of poignancy to the whole drama. It would be far more effective if a name was used instead of 'that poor mouse'. Be that as it may, once more the telephone rang. This time it was the chairman of the Johannesburg Society for the Prevention of Cruelty to Animals (SPCA)!

They were going to prosecute me! The animal lovers throughout our fair city demanded my blood for the callous brute that I was. I hastened to the offices and confronted the chairman. I had the little corpse in my wife's freezer and the post-mortem revealed heart failure, probably brought on by the loud bang when the motor ignited. I handed this report to the gentleman and he took a long time reading it. At last he put it down on his desk and looked at me in a most unfriendly way.

'If you pay a fine as admission of guilt we are prepared to drop the prosecution,' he said in a monotone. I was not prepared to do any such thing as admitting that I had, in any way, been cruel to any animal, let alone a mouse.

For years I had been treading on eggs with the Inspector of Explosives and now to admit to these charges would certainly sound the death knell to my research. I was not going to throw whatever small favour I might have had with the inspector down the drain on this ridiculous accusation.

This time I fixed the chairman with a steady gaze and told him that I would see him in court and that he had better bring bona fide witnesses as I had the unadulterated proof on film, movie film, of the launch. If any sane person could point out cruelty I would gladly accept the court's finding. I then asked him a pertinent question. If his wife told him that a mouse was in the kitchen, how would he see to its demise? By way of a cheap mousetrap, a device that had been in use for goodness knows how long. The number of mice that this simple but effective killing machine had dispensed with was anybody's guess. Also, if the trap was sprung and it did not strike the mouse a fatal blow but, say, broke the rodent's back, it would remain in pain, until some kind soul put it out of its misery!

No, Sir, I had not been cruel, I had attempted to recover it safely and unfortunately had failed. I had put my faith in an expensive machine in an attempt to further our knowledge of the unknown and although the mouse had died, more knowledge about the technology of rocket flight had been gained. He thought about what I had said and dismissed me with: 'We will be in contact.'

I left the office feeling anger at the manner in which I had been treated. It would have been laughable if it was not so narrow-minded! For the next week my name was heard on every radio newscast and I grew more frustrated daily. I was privy to a remark that was so unbelievable that I went once more to the office of the SPCA. In some circles people thought that I had stuffed the mouse inside the combustion chamber! Being half suffocated, it was then incinerated when the motor ignited!

I took the movie of the launch plus a projector with me. On seeing this, the chairman apologised and told me that the charge against me would be dropped. I understood the predicament he was in as his organisation depended on public funding and when they shouted they expected action! At the end of this meeting we shook hands and I departed a much-relieved man. The chairman notified the press as to the society's decision and the next day the daily newspaper carried an article under the heading: 'Mouse rocket-man not cruel'.

Redemption at last! There were a few parting shots and the matter was forgotten. Life goes on, but the anxiety and worry are left behind in the memory to come back and haunt one.

I had been busy with the design of a three-stage rocket prior to the mouse débâcle so I ploughed my energies back into completing it. The year 1960 had just begun so I figured the new year should start with a brand-new project. I had not attempted to assemble three rockets in tandem before, so here was a bright opportunity. Bearing in mind what had been discussed before with regard to mass ratios, this design was not all that straight-forward. All three stages had to be in proportion to one another, that is, the structure to fuel ratio. The first stage had to be the muscle to lift the other two, so it had to be the most powerful. The second stage was slightly smaller, but strong enough to push little brother to 3 048 m. The three stages telescoped into one another, being separated as each stage ignited, leaving the smallest stage to complete the final journey upward.

Put that way, it sounded simple. It was not. The biggest hurdle to overcome was the precise ignition of each motor.

A common misconception, even today, about multi-stage rockets concerns the point when the lower stages release from the upper. The general belief is that each stage carries the others to its maximum altitude then drops off, leaving the rest to carry on. The truth is that each stage is discarded when it has reached its maximum *velocity*, thereby enabling the final stage to reach its maximum velocity. At the moment of burnout a rocket immediately begins to slow down as *air drag* and *gravity* start to decelerate the vehicle. If ignition of the next stage is delayed by a few seconds, it will not have received the maximum boost from the previous stage owing to the whole rocket slowing down. If fired at the right time, the velocities may be double that of the previous stage. The crucial design criteria!

This was and is the method modern rockets use to reach orbital velocity, and yes, escape velocity that is required to reach the moon. The huge Saturn V rocket that carried humans to the moon in the late sixties was an enormous three-stage rocket. No single-stage vehicle is yet capable of attaining a velocity of 11 km/s, the velocity necessary to break the bonds of Earth's gravity. This is why the multi-stage rocket holds centre-stage as far as the exploration of space is concerned.

My finished rocket stood 1,94 m high and had a loaded weight of 40 kg. The ignition sequence was electronic in an attempt to avoid unreliable mechanical systems that had a bad track record. All in all, this rocket was the most ambi-

See pages 94 and 120 for enlarged versions

tious that we had undertaken and our hopes were high. I was out to send this one way past the 1,6 km mark, which still stood as the record altitude.

I had invited a few of the press to attend as well as the officials. The press accepted, but the officials gave no response. The setting-up of this rocket was a long, time-consuming exercise in caution and precision as there were three sections to contend with. A mistake at any one stage would wash the whole shoot down the wazoo! Care was the operative word as we worked toward an early morning lift-off. The rocket crew had slept at the site, having laboured into the early hours. With the sun brightening the eastern sky, the final items were attended to and the launch was on!

The countdown went without a hold and we were into the last seconds.

The electronic fire control on board the rocket was tested and cleared. This was the last check on the list as the time ticked to the pressing of the firing button.

'Ignition.'

The first-stage motor blasted into life, thick white smoke belching out in all directions! The slim yellow rocket leapt into the air spitting flame and smoke. It appeared to jump into the air and stop, growing smaller as it surged upward.

Because our position at the control panel was relatively close to the pad, the effect of the rocket's journey into the sky could not be fully appreciated. This was the reason that the ascending vehicle appeared to stop and grow smaller, the angle we had to tilt our heads back to see it, virtually presented a 'tail only' view! High above, the second stage ignited, blowing out a spectacular cloud of white smoke. Jack captured this split-second once-in-a-lifetime picture on film.

The first stage was seen to drop away from the rest of the speeding rocket and begin the long fall back to ground. It left a thin trail of smoke, growing ever longer, stark against the blue of the sky. But all was not well with the second and third stages' progress! The flight angle was not as it should

See page 116
for enlarged version

have been. The tower had been tilted at 2° off the vertical for the launching but the rocket was clearly not aligned to that angle. If anything, it was now travelling at more like 10° downrange!

The maximum was reached and the tiny speck trailing a thin vapour trail arced over. The third stage had not ignited! The two stages, still coupled together, plummeted like the proverbial bomb.

It struck the earth, too far away for any sound to be heard. At the control panel all was silent. In the distance a warbler whistled its merry tune, adding an unreal melody to the drama that had just played itself out. The post-launch procedures were carried out in what appeared to be a sound of mourning after the eerie whistling sound of the descending rocket – a whistling sound that was abruptly silenced.

The flight officer's voice broke the silence: 'Longest flight time we have had. Should be a good one!'

The tracking information confirmed the flight's prediction. I rounded the figure down to an official 2 290 m, our highest to date! An unpleasant task awaited me. The rocket had thrust deep into the ground with the still-fuelled third stage just sticking above ground level and the spent second stage still joined. It looked like some bizarre flower growing in the most unlikely place! I surveyed the situation and saw that the second stage could be removed from the tail of the final stage. This now afforded me the opportunity to insert a new igniter into the fuelled section and connect it to the firing panel, which had been brought closer.

With all personnel at a safe distance, the cable was joined to both rocket and panel. A warning was given and after a pause of 30 seconds, the firing button was pushed. The motor blasted its starting note and continued to throw fire and smoke upward into the air. At burnout, again a deathly hush descended over the range, disturbed only by the mocking sound of birdcalls. A potentially dangerous situation had occurred in two instances: the bottom stage being blown off at an angle, bringing it close to where the control point was; and the slight delay in second-stage ignition resulting in an angled flight.

This meant that the predicted point of impact was further away than was calculated. If the angle had been greater, this point could have been much further away. Although we had miles of clear area downrange, I did not want to experience a rocket flight that was not controllable! And why had the electronic sequence circuit failed? This was the cause of the wayward flight. On viewing the movie of the launch, the answer stared at me in the face! On ignition, the rocket left the pad in a fraction of a second, but when viewed in slow motion another story was evident. For a fleeting moment before lift-off, the entire rocket was enveloped in the exhaust flame. Any exposed soldered joints or electrical wiring was subjected to the intense heat of the motor and ionised particles in the exhaust. Result? Disaster! Everything in that environment would be affected. This problem could be easily solved but what about the other?

The event of the rocket deviating from its pre-set course was a major concern and therefore I had to come up with a major solution. The solution was to install a radio receiver in the rocket so that when the correct signal was transmitted to it, a relay would close and a destruct package would be activated. The rocket would then be broken at its centre of gravity, effectively destroying its aerodynamics and causing it to tumble. In this mode the now imposed high g forces would rapidly bring an end to further flight.

Who spoke about gremlins? I was beset with them! Each new rocket brought with it new hurdles to overcome. The challenges were endless but ones that would be conquered. More sophisticated rockets were being fired and each firing showed me my shortcomings. It was like winning a game of chess. You did not learn anything from it, particularly if your opponent made a silly move. But if you played a good game and were beaten by a nefarious, subtle move, you would learn never to be caught like that again!

So it was with resolute determination that I approached each new game on the drawing-board and each move of the cursor of my slide-rule. Always trying to see that little bit further ahead and precluding another gremlin. A 2,5 m two-stage rocket was taking shape with all of the foregoing modifications included. If ever a design would meet the requirements intended for it, here was the one! It was really something that was a breakaway from previous versions. All electrical equipment was protected from any heat-generating components and the radio receiver was protected in a sealed capsule. The aerial was taped to the wooden nosecone at the front, far away from the motor exhausts. I had changed the colours to white with black lettering and black spin markings. I had always opted for four fins, never mind if arrows had flown straight with three fletching feathers for centuries, I had had good results with four and I stayed with them.

A new condition was handed to me by the inspector when I gave him the design of this rocket for permission to launch.

In view of the amount of propellant to be used and the chemicals involved, he decreed that only one person would be allowed to fuel the motors and that person *must* wear a complete asbestos suit as worn by fire-fighting personnel! Here was a new safety requirement that, although expensive, made a lot of sense. Some years back I had accidentally been burnt by a small amount of experimental fuel I was mixing. It burnt my right bicep and I carry the scar still. I was all in favour of any logical suggestion as far as safety was concerned so I informed the man that his condition would be acceded to. But where does one obtain such an item? That's right. The Fire Department!

The Chief Fire Officer of Johannesburg listened to my story and request in what could only be described as shock! Again it appeared as if my sanity was in doubt! Asbestos suit? Did I have the slightest idea as to the cost of one of these units? No? Did I think that they were easy to come by? No! Then how the @#* could I come and ask to borrow one on the odd Friday and return it on Monday morning? I explained that without the suit my project would be scrapped and if that happened, a lot of work would go down the tubes.

'Aren't you responsible for shooting a mouse into the air a little while back?' he enquired.

'Yes, Sir. It was a scientific experiment into the feasibility of subjecting the mouse to an acceleration force of 5 *g* straight up and return it safely back to ground. The parachute opened prematurely and was torn off. The mouse died of heart failure and I was in big trouble with my wife owing to my keeping the corpse in the freezer until the enquiry was over.' My voice sounded strange as I relived those anxious times over again.

He suddenly smiled and told me to collect a suit at the stores across the road from his office. I thanked him profusely and rapidly departed before he changed his mind! An asbestos suit is not a small item. I found this out as I signed an official document for an official-looking large cardboard box. It was not heavy, just bulky, and barely fitted onto the back seat of my car. I had my suit! Not permanently, of course, but on loan.

The following week saw the final preparations completed for the upcoming launch. The rocket was set to reach over a mile carrying the new payload consisting of the radio and destruct package. Without this additional weight, the altitude would have been just on double, but I accepted this in favour of the safety that these items offered. As launch day drew near, I felt a strange feeling of excitement I had not felt before. I could not understand why and I put it down to the new rocket and my having to wear that clumsy suit!

The advantage of the new launch site was that we could cover the launch tower as protection from the weather but leave it assembled in situ. A great time saver! The rocket crew had most of the work fuelling, checking out the various systems and setting up. The control men had their time readying the panel and the associated circuitry. This now comprised the additional wiring for the destruct transmitter, the high-voltage charging system for the tracking cloud and the firing control. We really had travelled a long and arduous road. Nerve-racking at times, frustrating at others, but in the end, rewarding. I was criticised in the press by some kind soul asking why I was doing something that was being undertaken by much larger nations elsewhere in the world and *they* had already put satellites into orbit. My published response to him was polite and to the point. Why should we not achieve knowledge or break-through for ourselves instead of relying on other nations to supply the answers and act as Big Brother? I ended with Pascal's maxim 'that experiments are the true teachers that one must follow in physics'. I heard no more.

It was time to don the shiny, aluminium-covered asbestos suit! One advantage was that it dispensed with the gas mask and the sweat that ran into my eyes. Putting the suit on was no easy task. And with the metal overshoes, walking was a clumsy affair, to say the least! When the all-covering hood went over my head I felt claustrophobic! Movement was not that easy and I tried a few gliding dance steps to the delight of my crew! I was fortunate enough not to end up on my behind! Games over, now to the business at hand.

Although it was early morning with the usual chill in the air, inside that thing I was sweating like never before. Every

See pages 93 and 114 for enlarged versions

move was an effort and because most of the work had to be performed at or below waist level, the suit was most unco-operative. It had its own ideas about bending in the middle! Finally after what seemed like an eternity, the motors were fuelled and the rocket was safely in the tower. I made the final connections and laboured my way to the control point. LET ME OUT!

The cool breeze on my face was like being hit with a dash of ice water and all was back to normal. The electronic boys were running systems checks and I squatted beside the firing officer at the panel.

'Everything is checking out OK and we should be launching on time,' he remarked.

'On time' was 08:15 – the time that we had set to fire more than a week ago to comply with the Department of Civil Aviation's demand to have the rocket back on the ground before 08:30. A new condition so as not to accidentally shoot down an airliner on the way to the Cape! Fair enough. The countdown had begun and the concentration was a tangible thing that could be felt by all of us.

The flight safety officer began reading the events from his clipboard in a voice we had all come to recognise and somehow it brought home the fact that this was no game. We were trying to capture the secrets of an entirely new scientific field. Each one of the members of the SARRG was as dedicated to this research as I. We were out to win.

'Power?'
'On.'
'Master?'
'On.'
'Heater?'
'On.'
'Radio?'
'Standby.' The sequence and affirmatives came swift and positive.
'Rocket clear.'
'... 3, 2, 1. Ignition!'
The firing button was deftly pushed.
'Plus 1, 2, 3, 4, 5. Cut-off!'

If the motor had not started after five seconds, the control panel was immediately shut down and the range was declared unsafe. I had written this into the launch procedure just in case, and here the situation had arisen! My worst worry had come to visit! A wait of one minute was stipulated and when this time had elapsed, it was back into the suit for me. When the firing button had been pushed, a puff of smoke was seen to exit from the first-stage nozzle and nothing more. The igniter had fired but no ignition of the motor had occurred. A first for us. I removed the safety key from the panel and donned the head-gear once more.

After nearly one hundred static firings and more than five hundred flights, a misfire on the pad! I walked with a clumsy sashay to the rocket. The spent igniter was lying on the ground and it was the only sign that the control panel had worked. I placed a screwdriver up the nozzle and scraped a little propellant out. This I took some distance away and attempted to ignite it with a match. Nothing! The first stage was loaded with contaminated oxidiser!

I called one of the rocket crew to help me disassemble the two stages from each other and after he had gone back to the control point I proceeded to empty the fuel contents into a suitable container for later disposal. Back into the suit and another batch was mixed. I tested this in the same manner and was relieved to see it crackle into life! Fuel the rocket and let the show carry on! The countdown was resumed at T minus 30 seconds and went down to zero with no hitches.

I had discarded the suit as soon as I arrived back at the control point and stood behind the firing officer. All the various panel lights were glowing at the correct time, indicating systems operation was proceeding smoothly. I replaced the key into the panel as the count reached 10 seconds. The ignition command was initiated at the precise time and the red indicator lit up like the one-eyed Cyclops. Ignition!

The rocket roared off the pad and streaked upwards, a second gout of flame and smoke burst forth as the second stage announced its new-found power. This fellow was really moving! Up, up it continued until finally a snow-white cloud suddenly appeared in the sky. The rocket had turned over and was headed down. I signalled to the firing officer and he pushed the button with a red ring around it. The destruct transmitter! Immediately the falling rocket broke in two amidst a burst of smoke, followed by a dull 'boom' a few seconds later. The two pieces tumbled to the ground and landed with a barely audible thud. It had worked! We had control of a flying vehicle and the confidence could be felt by all.

The tracking had been excellent and the final altitude was 2,4 km!

Slightly higher than the ill-fated three-stage rocket and another record in the pocket. We were really getting better! The press had been there to witness this and instead of a picture of the rocket, a half-plate photograph was splashed on the front page showing me in that infernal suit, looking like something from outer space! Maybe it was not too short of the mark after all!

Chapter 4

The fire that thunders

The years I had spent with the design, making and launching of solid fuel rockets had run parallel with my desire to enter into the arena of liquid fuel motors. This type of motor was more complicated than the units I had built over the years. By the very nature of the propellants that were necessary it also made the handling more difficult and therefore more dangerous. As was pointed out in an earlier chapter, one of the fuels had to contain oxygen to sustain combustion. In a solid fuel motor, this oxygen was obtained from chemicals such as potassium nitrate, potassium perchlorate or any one of the other perchlorates. One of these mixed with an appropriate organic chemical produced the desired reaction when ignited. But these compounds carried their own hazards as well, and could never be treated without the utmost respect

However, the oxygen component that I proposed to use in the liquid rocket motor is the widely used liquid oxygen. This oxidiser boils at −183 °C and presents difficulties in maintaining it in its liquid state. Avoidance with the bare human skin is to be observed at all costs. The very low temperature of this substance may be demonstrated by dipping a length of garden hose into it and then gently tapping it on a solid surface. The hose shatters as if made of glass! What then would our skin suffer? For all of its undesirable properties, liquid oxygen remains the most common oxidiser in use today. It has a pale blue colour and spews a dense white vapour as it rapidly boils away at atmospheric temperature.

It is generally the last fuel to be tanked into the rocket and it is responsible for the white vapour swirling around the vehicle prior to lift-off. A dense ice formation on the outside skin of the rocket is seen and it is this ice that breaks away and falls as the motors ignite and take off occurs. This ice provides good insulation in maintaining the oxygen in its liquid state within the on-board tank Liquid oxygen is one of the safest with regard to bumping or jarring and does not react to hydrocarbons unless heated, that is, ignited. On ignition, furious combustion takes place.

Other oxidisers are high-strength hydrogen peroxide, red fuming nitric acid, and nitrogen tetroxide. The former two are highly hazardous and are used only in special applications. Red fuming nitric acid combining with hydrazine produces spontaneous combustion! If all of the foregoing are taken into account, one must surely ask: 'Why the risk?' In fact, what we are dealing with is a controlled explosion!

As technology has advanced, the application to specific needs has been solved and much of the inherent danger of these chemicals has been overcome. (A classic example of this is atomic fission. We have learnt to harness the awesome power to meet peaceful requirements to our advantage.) Accidents relating to the use of chemicals are by and large relatively rare in rocketry today. The precautions in vogue at present have evolved from more than 80 years of perfecting techniques in the handling of these chemicals.

The compounds used to react with these oxidisers are far more docile, but being highly flammable they too must be respected. Ethyl alcohol, petrol, refined paraffin and liquid hydrogen have been successfully used. This produces a far more controllable reaction when mixed in the correct proportion with an oxidising agent in a specially designed combustion chamber and ignited. This type of rocket motor was used to land a man on the moon in 1969 and is the workhorse of all odysseys to our neighbouring planets and satellite launchings. At the time of writing, 102 launchings of the space shuttle have taken place with one disaster. A 1% failure rate! Nobody ever said that exploration into any new technology would be easy.

My delving into the intricacies of liquid fuel as a means of propulsion was a natural extension of the solid rocket. I wanted a longer operating time with regard to the thrust period. The available solid chemicals restricted me at that time and therefore the total impulse from the motor limited the altitude. (The total impulse was calculated by multiplying the burning time by the average thrust of the motor.) Total impulse can be increased by building a bigger rocket. The problem was rather that the specific impulse (= total impulse/propellant mass) was low, that is, the propellant was inefficient. Liquid propellants provide a higher specific impulse and this avenue needed to be explored. There was a lot more hardware to this motor and here are the main components:

- a tank to store the liquid oxygen
- a tank to store the hydrocarbon fuel
- a combustion chamber to burn the fuel mix plus the necessary cooling system for same
- the necessary plumbing to lead the liquids to the chamber
- the valves to admit the fuel mix into the chamber
- injectors to mix, atomise and vaporise the propellants
- a compressed gas tank to force the fuels into the chamber
- finally, an ignition system that would ignite the flammable mixture in the chamber.

A formidable list of equipment in any language! The layout of all these pieces was not that difficult, if you considered that the tanks and the chamber had to be assembled in tandem. The compressed gas at the top followed by the fuel tanks with the combustion chamber at the bottom. The plumbing connected the compressed gas to the fuel tanks, which in turn connected to the chamber at the bottom! Simple? Not on your life! I played with a mock-up of these bits and pieces for months, attempting to achieve the best configuration for the pipes and valves. Once I had a semblance of the pipe routing and valve positions sorted out, I concentrated on the combustion chamber parameters, for this was the very heart of the entire project. The anticipated temperatures that the motor would operate at were between 2 800 and 3 500 °C! This extreme range of heat was sufficient to melt any metal that I knew of. Cooling the chamber was imperative if it was to operate for any time over one or a maximum of two seconds.

I had set the assembled tankage and piping in a vertical framework with the chamber offset to one side. In this position the nozzle was pointed slightly at right angles to the vertical. This was my standard practice for static testing any rocket. If the motor tore loose from its mooring, it could not take flight and land where it might do damage. A sand bunker was directly in front of the motor to arrest any unwanted movement away from the test rig.

This method had proved itself in the past and, happy to say, I never had to retrieve any part of a rocket out of that bunker! The chamber as finally conceived was 29 cm long and 10 cm in diameter. The two injectors were welded into the head-end of the chamber and the propellant pipes connected to these. The chamber was formed out of 3 mm steel. This unit was not intended to fly but rather to act as test bed for the one that we hoped would hopefully take to the sky in the future.

Another innovation was that gaseous oxygen would be fed into the appropriate tank and not the freezing liquid. This was for cost savings during the experimental work. I decided to use an inert gas to dispel the petrol from its tank and nitrogen was selected. The igniter was an electrically fired, pyrotechnic 'Catherine wheel' mounted on a wooden dowel and inserted up the exhaust nozzle. Once ignited, the wheel would spin round, distributing flame in all directions inside the chamber. The starting valves were wire-operated from the safety of our dugout, some ten metres away.

About eight years of work were now ready to show how far along the liquid rocket motor road I had travelled. Much thought, time and energy had brought me this far but was it far enough? I was only seconds from finding out. My first liquid fuel motor was on the test stand and was ready to be proven. Many months of planning and building saw the finished unit standing in front of us, now ready.

The nitrogen tank was at half-pressure for this run and only a half litre of petrol was loaded. The oxygen cylinder was in the dugout with me and a long hose connected it to the tank out on the stand. A twist on the lever on the cylinder sent a hiss into the tank. I pressurised it to 14 atm and left it at that. We kept silent and listened for any telltale hiss that meant a leak somewhere in the valves or joints. Nothing was heard except the whistling of beetles and the odd bird-call.

The sequence of events was worked out weeks before and we followed them now:

'Ignition.' The igniter sputtered into life, smoke streamed from the nozzle.

'Fuel.' A long tongue of orange flame leaped into the smoke as the petrol entered.

'Oxygen.'

A loud ear-splitting 'Crack!' followed immediately! I shut off the oxygen supply and watched, helpless, as the last of the petrol burnt away. Out on the stand there was just devastation. A piece of chamber hung by one pipe and the starting valves were bent out of line owing to the piping being at all angles. We had not had a start – we had had an explosion! I had gone about the sequence all wrong! It hit as if by a physical blow!

I had let the petrol into the chamber first! There being no oxygen present to ignite it, a dangerous amount of the liquid had collected. When the oxygen entered a split second later and ignited the excess petrol, a volume of propellant exploded, far exceeding the design pressure of the chamber. I had experienced my first and only 'hard start'.

Of course, the oxygen must enter first so that as the fuel entered, combustion was initiated immediately. Another hard and bitter lesson!

The next test run saw the system restored to its original form and once more we waited behind the safety of our dugout. The commands were given and this time I prayed that I had got it right. The run was scheduled for three seconds and when the petrol followed the oxygen a loud 'whoomp' blasted out. It had started!

My elation was short-lived, as the short spear of flame being emitted from the nozzle grew smaller as another appeared shooting upward. In less than two seconds the entry into the nozzle began melting away. I shut off the oxygen supply and looked at another smoking ruin. The silence seemed to sing in my ears after the noise of the rocket. A sound like I had never heard before. Something like a motor bike without baffles together with the roar of a huge waterfall. The combination was incredible!

My initiation into the élite arena of the fire that thunders had been endured and I had survived to fire another day.

On inspecting the melted chamber, I stared at the gaping hole at the beginning of the convergence section of the nozzle. It was at this point that the combustion gases started to accelerate to supersonic speed as they converged into the throat. Once through this area the gases expanded and accelerated further in the wider divergence section of the nozzle, adding more to the forward thrust.

The hole was at the position where the most heat would be generated and the intensity had reached beyond the melting point of the steel, causing the blow-out. Some form of cooling was imperative. Not only at the nozzle area, but the complete chamber had to be protected from the heat transfer intensity. Making the chamber thicker would succeed only in producing a heavier unit totally unsuitable for flight. This might be a temporary solution but ultimately it too would succumb to melting after a slightly longer burn.

The solution to this proved to be fairly simple on paper but a difficult one to make. It consisted of a cooling jacket entirely encompassing the chamber and injector head. The petrol entered the jacket at the bottom of the nozzle and was circulated around the chamber to be injected at the head. The oxygen injector was a separate unit, angled to the fuel injector, which had its inlet in the circulating cooling fuel. The two injector outlets faced each other in the head, allowing the propellants to impinge on each other, thus mixing together. This mixing produced atomisation and vaporisation essential for complete combustion.

The finished product could be likened to a can containing the chamber. The petrol inlet into the can was at the bottom and the whole was sealed at the top with the oxygen inlet pipe sticking up. When the fuel system was pressurised and the valve opened, petrol rushed into the bottom of the can and swiftly circulated around the chamber. It was then forced through the injector and into the chamber at the head. This swirling, continuous flow of cool petrol transferred the heat away from the metal walls, effectively preventing any hot-spots where overheating might occur.

This was the theory and the new liquid fuel motor was built according to this. The tanks were the same, as were the valves and compressed gas cylinder, but the chamber was mounted vertically. The test stand had to be modified somewhat to accept the new configuration and the hold-down bolts increased. In essence the procedures were the same and the oxygen cylinder was in the dugout with me as before. I had calculated a firing time of five seconds and accordingly the correct amount of petrol was loaded into the tank.

A valve was opened and nitrogen hissed into the petrol tank. I released oxygen from the cylinder and pressurised that system. The igniter inside the chamber sent sparks pouring out of the nozzle. Our eyes were fixed on the base of the motor. I opened the oxygen valve and the sparks grew brighter as they now cascaded downwards. The petrol was released from its tank and a loud 'pop' was heard. (This was caused by the petrol fumes entering the chamber.) As the petrol entered into the chamber, the loud 'whoomp' we had heard once before smote our ears! The motor had started smoothly and was roaring out its power in a short orange flame. Clouds of dust were being blown into the air as the din continued. I watched the flame in fascination, expecting it to burst from somewhere else. But still the roaring continued!

A splutter, another one, and the flame and sound died suddenly. The petrol tank was empty and I hastily closed the oxygen valve. The noise and dust had seemed to go on endlessly! My ears were ringing and I was aware of an unusual silence. Deafness? No. Sound came back slowly and I felt a hand on my head:

'You really did it that time, my boy.' I did not recognise the voice. It was the firing officer.

'The motor ran for nearly six seconds and the whole thing is still standing!'

I had designed it to produce 50 kg of thrust and as I sat looking at it, strange creaking noises were heard as the hot motor began to cool and this made the firing crew look steadily at it – almost expecting it to end its symphony with a final explosion of sound! It never came. We stayed in the dugout making the odd comment, not willing to end the experience of witnessing the first successful firing of a liquid fuel rocket motor in South Africa!

The date was 6 October 1959. A fine day to remember.

The first firing of the motor using liquid oxygen was another story! I had been schooled by the manufacturers as to the handling of the liquid before they would release the vacuum flask and its contents to me. 'The pipe from the flask to the rocket must first be cooled for two minutes before the oxygen is fed into the tank.' 'Yes, Sir.' 'Remember not to exceed the time that the liquid is in the tank according to your calculations. The vaporisation may exceed your tank pressure.' 'Yes, Sir.'

I was no hero and I appreciated every word of advice given to me by the kind sponsor. It was not a chemical to be taken lightly and these gentlemen had gone out of their way to assist me with the oxygen and teaching. To them my appreciation. (No publicity, please.)

The day for the first static run was set and the complete motor was disassembled and thoroughly inspected. This burn was to be the one with all liquid fuels! No oversight was going to rob me of this one performing as the last run had. The joy and excitement still pervaded my very soul! This one was going to be even greater as it was the motor scheduled for the first flight. If this operated as planned it would be shelved for a month, as we had been committed to a solid rocket shoot in ten days' time. (One of the Saxon series.)

The building of the airframe for the liquid rocket was far more involved than any solid fuel rocket we had built. The outer shell was fabricated from aluminium sheet.

The inner struts were aluminium angle. These struts were the supports to fix the whole motor and tanks firmly in place. A cross-brace near the tail was the attachment for the motor itself and would also transmit the thrust to the airframe. A panel was cut into the side of the shell exactly where the top of the oxygen tank was located. A pressure gauge was fitted to the top of this tank and had to be visible to me in the bunker. More about this gauge later. Four delta-shaped fins were riveted at the tail section. These were spaced 90° apart. The aluminium nosecone completed the external appearance. Again the colours were white with black stripes and lettering. This rocket proudly sported the model name 'Saxon Alpha Mk 2'. The rocket was constructed in such a manner that the motor and the tanks could easily be inserted or removed. All fixing screws inside the airframe were easily accessible to accommodate the installation of the propulsion unit. Just behind the nosecone, the forward compartment of electronics was located. If the solid fuel rockets lifted off at great speed, this one would appear to stagger into the air!

My hopes were high on the afternoon of the final static run prior to launch of the liquid fuel rocket. The test stand was set up exactly as before: the compressed nitrogen tank at the top and the motor at the bottom. To keep the weight distribution correct it was found that the petrol tank had to be above the oxygen tank. The mixture ratio of oxygen to fuel was close to 3 to 1, therefore the oxygen (O_2) tank was bigger and heavier. By this placement I was able to keep the centre of gravity of the fully fuelled vehicle forward of the centre of aerodynamic pressure. This configuration was vital if the rocket was to fly stable along a vertical path.

All the preparations had been finalised when a light van arrived at the test site. The liquid oxygen had been delivered! The technician showed us the loading procedure from the flask into the tank. We were surprised to see how rapidly ice formed on the outside of the tank and piping. A vent at the top of the tank was left open to allow the evaporated gas to escape. This hissed into the air leaving a white shroud swirling around the stand. Suddenly a pale blue liquid spouted from the vent. The tank was full.

The vent was sealed and we retired to the dugout. The starting valves were replaced with electrically operated burst diaphragms, a great improvement over the wire-opened jobs. When the switch was closed, an electric current activated a small explosive, which drove a thin spike through the seal in the pipe-line. The pipes now being opened, the propellants were allowed to flow.

Earlier I mentioned a pressure gauge on the oxygen tank.

I watched this gauge through a pair of binoculars, waiting for the needle to reach 20,4 atm. This was the desired pressure to feed the O_2 into the chamber and to start the motor. One switch on the panel would fire the igniter, the nitrogen tank's diaphragm and also that of the O_2 tank. The O_2 tank being closer to the chamber allowed the O_2 to arrive first, precluding a 'hard start'. The needle edged closer to the red stripe on the gauge as the pressure increased in the O_2 tank. (I told you it was heavier!) It reached the mark and I called: 'Ignition.'

First smoke issued from the nozzle followed by the 'whoomp', signalling a perfect start! The motor roared its demand to be released in clouds of dust and a bright orange flame stabbed at the ground. The technician who delivered the O_2 flask cupped his hands over his ears and watched with wide eyes as the burn progressed. The thunder seemed to roll across the open range, putting birds to flight. The cut-off came suddenly. A great silence followed. The last remnants of petrol burned out in a quiet sheet of flame. The stand and motor seemed to shimmer within the whirls of thin smoke being emitted from the scorching hot motor. The time of burning was as close to five seconds as anyone could want. Our liquid fuel rocket was ready to fly!

The gentleman secured the vacuum flask in his van and spoke for the first time since the test: 'That is one thing I would not care to be on the wrong side of. It was really great to be here and see it.' He was truly impressed and stayed with us while the equipment cooled down. We took this time to have a cup of coffee from our thermos flasks that we had brought with us. What a burn!

We packed the motor and its ancillary equipment into our motor cars and prepared to leave the site. I sat quietly in my car and watched the dust of the departing vehicles in the distance. The test stand was a silhouette against a most beautiful sunset. I was happy and it was a perfect end to years of work. As the darkness deepened, my thoughts returned to the very beginning of this whole opera and the many acts that had been played. And as yet the 'fat lady' had still to sing the final aria. I switched on the ignition of my car and started the engine. A wave of wellbeing swept over me as I concentrated on the headlights in front of me. I was anxious to get home and tell my wife.

The launching of this rocket was not notified to any of the news media. I wanted to keep it quiet, as the memory of the mouse débâcle was most unpleasant. I certainly did not wish to experience that hullabaloo again. This was an unproven technology here in South Africa and I wanted to display its potential in relative secrecy first. The date was set to take place a month after that successful static run. The building of the airframe was almost complete and a few minor modifications were required to the electrics behind the nosecone. The rocket was 1,7 m tall and weighed 30 kg when fully fuelled. It was 15 cm in diameter.

Because of the relatively slow lift-off, special attachments were fitted to hold it to the tower at the initial rise. These guides were to ensure that the rocket slid out of the tower smoothly and adhered to the pre-set flight angle. As with the static runs, all the pre-launch preparations were carried out except for the loading of the liquid oxygen. This was the last operation before the countdown went to the final sequence. The arrival of the van was anxiously awaited and I was relieved to see the dust rising along the sand road.

The van pulled up next to the tower and the same technician as before emerged. The piping from the flask was connected to the internal tank and the cooling process began. Ice started to form almost immediately and the blue liquid began to flow into the tank, sending white plumes of vapour into the air. This venting continued, and the condensation on the outside of the airframe soon turned into ice as well. With a spurt, liquid shot out of the vent. The flow into the rocket was shut off and the vent closed. The crew were at their posts when we hurried in from the pad, parking the van behind the dugout. I focused my binoculars on the pressure gauge inside the rocket.

My heart pounded as the needle steadily swung up, indicating the pressure rise inside the O_2 tank. The systems checks had been completed and the silence seemed unbearable during those final seconds, waiting for me to give the irreversible command. The needle now edged to the red line and as it reached it, I called:

'Ignition!'

There was a slight pause as the igniter flared and the burst diaphragms ruptured, unlike the instantaneous take-off of solid fuel motors. At 100 m

See page 107 for enlarged version

distance the sound of the motor-start was still unnerving! The smoke from the igniter formed a small inverted mushroom on the ground and the rocket moved slowly up the guiderails. It cleared the tower and rose faster, gathering speed rapidly as weight decreased in the fuel tanks. That sound like thunder beat down on us from above, splitting the very air with its fury.

The flame from the exhaust could hardly be seen against the pale blue sky. Suddenly the sound ceased – the motor had stopped prematurely! The rocket coasted at a leisurely pace upwards. The white and black spire of the airframe could clearly be seen from the ground as it stopped and seemed to pause before it slid backwards towards the earth. As it increased in speed, the nose pitched over and the rocket hurtled down. A white streamer of vapour behind it quickly dispersed in the gentle wind.

It slammed into the ground with the sound of buckling metal within 25 m of the tower. The crew at the control point witnessed the sound and sight. It was not a pretty sight. The once-statuesque form now lay in a smouldering wreck. Something had gone horribly wrong for the motor to cut out halfway through the burn. I had not seen any warning signs that could have heralded an impending cut-off during the brief ascent. As always, nobody moved from his station as an unsafe situation could unfold.

We did not have long to wait before the unused petrol flowed from the broken tank and ignited in a bright flame in the oxygen-rich atmosphere around the fallen rocket.

The petrol soon burnt itself out, leaving a pall of dark smoke pointing skyward as if to say: 'In that direction.' The height reached was nearly 300 m.

At each firing, static or flight, a fire extinguisher was at hand as a precaution but it was not required. The rocket had landed in an area of hard soil and no burning of the environment took place. We had to wait for some time before things cooled down sufficiently for me to examine the wreckage and try to determine the cause. I squatted beside the blackened hulk and stared at the motor section. No obvious fault there. My eyes moved to the plumbing and tanks. The cause had to be in this section. And there it was! The vent in the O_2 tank had opened! The pressure had escaped from the tank. The motor had ceased to thrust owing to oxygen starvation in the chamber. The flame had extinguished! In fact, the extreme cold had cracked the seal and the vibration from the operating motor had done the rest. This seal had been replaced after every static run and had not shown any untoward sign of failure. We had seen this as a possible weak link and had taken steps to ensure reliable working – thus the replacements. However, when the gods fail to smile on you there is not much that can be done! In this case we had lost the most expensive rocket we had ever built.

It had not achieved sufficient height for the parachute to deploy. The mercury switch and its associated release charge had functioned as designed but the rocket had slid back precious metres thereby preventing the mercury from shorting the contacts in the switch. When the rocket had finally pointed nose down, it was too close to the ground and the 'chute started to deploy on impact.

The pieces of the destruction were gathered and taken to the nearest dumpsite and deposited. Not one bit of the entire structure was worth salvaging, not a valve, not the chamber or tanks. All were dented or bent beyond further use. We had attempted an ambitious flight and had failed because of an asbestos washer that cost a few cents. The altitude reached was far less than we had hoped for, but altitude was not the prime reason for the launch. It was to prove that a liquid fuel rocket could be designed and made to operate successfully as a prelude to a satellite launch attempt. In my heart I felt that we had accomplished the object.

The motor and its method of cooling had functioned as designed, and this in itself was a major plus, for at that time was no literature on how to make a rocket motor that burns liquid fuel. Once the basics of rocket propulsion were understood, it was the degree of complexity that divided the solid from the liquid fuel application. For both systems followed the same laws that Newton had established so long ago. These laws will forever be applicable for as long as people venture into outer space.

I shelved any further development of the motor for the time being as it had eaten a big hole in our resources and other plans were looming in the solid rocket area. The launching of a large two-stage rocket capable of attaining 3 000 m (10 000 ft) was being planned and this presented an opportunity to go another step forward in developing large solid propellant boosters. If a person could throw all that we had achieved into a boiling pot he would undoubtedly see the direction in which we were headed – outer space with a satellite in Earth orbit. There were military implications, of course, but these were not in my grand schemes of things. I certainly never wanted to be part of deliberate destruction of life or property with my rockets. The goal was, and would always be, for science.

Chapter 5

Events with a double bang

To reach the altitude required would entail a big rocket, much bigger than I had attempted. The first stage would have to give the upper stage a big push up the wazoo to impart to it the necessary velocity for it to add to. The ignition sequencing would have to be gauged very accurately to start the upper as soon as the lower stage had reached maximum velocity, that is, at burnout. The system that I had used on multi-stage vehicles in past firings had not proved all that reliable and, in truth, I was not comfortable about using it on such a large rocket where the *g* forces would be greater than ever before. To design a new system in time to meet the launch date when the whole crew would be available was pushing my needle into the red!

I had not finished the design of the first stage yet and already the staff of the factory that would make the beast were bellowing for the drawings! It seemed as if everyone was chasing my tail and I was running around in ever-decreasing circles! Sanity prevailed and the drawings were taking on more and more of a professional appearance instead of rough sketches intermingled with calculations. I wanted to reduce the drag induced by the frontal area to a minimum on the upper stage, as it would be the one to achieve the greatest velocity. To prove this, just point your finger out of the window of a moving car and then face the palm of your hand into the air rushing past the moving car! This is drag and it is this unwanted force that prevents a poorly designed rocket from reaching higher altitudes. This rocket was being prepared to reach where no other had been in South Africa.

The motor sections were made out of solid drawn steel pipes and had to withstand great internal pressures whilst the propellant was burning. The nozzles were turned out of cold, rolled steel, as were the forward bulkheads. The fins were of the clipped delta planform for the lower stage and the delta planform for the upper. The nosecone was conical. The telescopic joining of the stages was employed to facilitate strength and easy separation at staging as we had witnessed in past firings.

The forward section carried the usual electronics in the form of the destruct transmitter and the smoke emitting circuitry. When finally assembled the rocket stood 3,4 m tall and weighed 69 kg fully loaded. It wore the white and black colours that we had now adopted as our permanent visual marking. The sky in South Africa is normally a deep blue except on exceptionally hot days, when it takes on a shimmering lighter hue. With this type of background the rocket was conspicuous virtually throughout the ascent. Remember that old maxim: 'KISS'. (Keep it simple, stupid!) This I attempted to do throughout my rocket days as it was my experience that the complicated things had a horrible way of turning around and biting me!

The launch day drew closer and I again extended an invitation to the authorities. This time I got a 'Possibly'. In the meantime we had achieved a great breakthrough as far as a launch site was concerned! The tract of ground belonged to a private company, well known in South Africa, and it went on for miles in any direction. Here was the perfect area to launch for many years to come. It had everything and nothing!

The manager and I walked around the spot that I considered to be the best locale for the launch tower. From this point, the area downrange extended for nearly 14 km and was 8 km wide! I had shown him photographs of the previous range and our dugout that we used as protection. He was silent as he gazed about him.

'How about it if I had my chaps build a proper dugout with overhead protection? We could throw the sand from the hole in a long heap facing the tower and dig the pit deep enough so that your eyes would be at ground level. We could install lengths of down-pipes through the sand at ground level so your crew would not have to be exposed at all during the launch. They would be able to see clearly through the pipes. I have read about your work in the newspapers and also magazines, so I know that you are dead serious in your research. Also, that you never had a rocket come down on your control point. I would be happier if we put railway sleepers over the top of the pit to give you even more protection.'

I could hardly believe what I was hearing! A proper firing range! The entrance into the bunker would be at one side (the east side, it ran from east to west with the downrange to the south). On further discussion, a trench would be dug, extending from the bunker to the site of the tower. This would enable the electric cable to be buried underground, saving the time running it out and rolling it up at every shoot. The ends were to terminate in a waterproof container. The last suggestion concerned the launch pad and tower. It was decided to fix the pad permanently into the ground and to erect and take down the tower at each launch. This was to prevent the possibility of theft. This was not the case with the pad, its legs and main framework being cemented into the ground.

My newly acquired friend told me the time frame he required to complete the installation and I was only too happy to hear that it fell before the date that we had set to launch the large two-stage job. My colleagues were overjoyed to hear about this turn of events. A flagpole was offered, to fly a green flag when the range was safe, a yellow flag when fuelling was in progress and a red flag when the countdown began.

This was now a highly professional operation and to complete the site, a public address (PA) system was offered! All of us, at one time or another, visited the range and offered any assistance that might be required as the work was under way. It was quite amazing to witness the rapid transformation of this once empty piece of land!

On the Friday afternoon prior to the shoot, I went to the range and started to wire the control desks. A ledge had been provided on which to stand the equipment and I marvelled at the visibility that I had, looking through one of the observation pipes. The whole launch area was there before me 100 m away. As I stood there, I could hear the flight's voice, counting the seconds to ignition, and the whole scene took on one of being unreal. The silence, the smell of newly dug soil and the anxiety of the events that would take place on the Sunday pervaded my senses.

It was dark when I finally packed up and started the drive home. A kaleidoscope of the many past launchings ran through my memory. The elation and the abject disappointments were all part of that and of me, which went to make up me! On reaching home I showered and sat down to a most welcome meal. For a long time after supper I sat outside, staring at the white rocket pointing up at the bright stars. On Sunday we would bring it to life, but in that short time, would it behave as we wished? It was a giant, waiting to flex its muscles and we had seen to its birth.

The following morning I drove back to the range with all the equipment plus the rocket and its unmixed fuel. There was a sand road that led to the bunker and as I drove slowly along it, wild fowl took flight, screaming their objections to my presence – a presence that would upset them far more than my vehicle in the not-too-distant future! Two other cars were parked alongside the bunker. One I recognised but the other was unfamiliar.

The flagpole had been erected and a green flag flapped in the breeze. At the pole, three men stood in conversation and turned to look at me as I parked the car. One was my friend Gordon and the other two were from the press. They had got the date wrong and were under the impression that the firing was today. Ah, the science of communication, or was it the lack of? They left in a cloud of red dust.

Gordon had brought the flags and the pole, and had been starting to assemble the tower when the gentlemen from the press arrived. They had helped him to erect the pole and hoist the green flag, thinking that the launch had been not too far from being executed. He told them that it was scheduled for the following day. I never found out who had told them that the shot was to take place that day. Anyway, they appeared to be uninterested in the facts and drove off, no doubt to another assignment.

A raised red scar disfigured the short green scrub from the bunker to the pad where the electric cable was buried. I started at the bunker end and began to separate and join the coloured cables to their appropriate circuits that would be connected to the control panel the following day. It was pleasant work and the peace and cool breeze acted as a stimulus to me. Gordon was busy with the tower and needed no assistance.

The conversation between us was clearly audible over the 100 m that divided us owing to the ambient silence and the direction of the breeze. We were unaware of the rapid passing of time and we found ourselves sitting on the pad and reminiscing over the many diverse subjects we had in common, such as sport, cars and family. He had long finished with the tower and I was completing the last of the connections at the base of the pad. In a junction box where the cable terminated was a multi-contact relay that was operated from the control panel. This relay would distribute the power to the numerous electrical systems in the rocket. The appropriate batteries were also housed in this box, which was 5 m from the pad in a special gully. This was to protect it from the soaring exhaust when the rocket lifted off the pad. A metal cover closed the gully.

The crew had decided that instead of leaving their beds at some ghastly hour in the morning, they would sleep at the range the night before and therefore save time. This was the first of many nights spent under the stars. Now that the launch pad and wiring were completed, the hard part had been taken care of and the remaining items could be attended to in a decent manner! No more rush or panic! The day had sped by and I was surprised to see the dark shadows of the tower pointing long fingers to the east. Time for me to go home and supper. Gordon and I left together, our spirits buoyed up by our efforts of that day; into our cars and a leisurely ride to our homes.

The next day saw the final adjustments and preparations completed. The two stages of the rocket were not joined together as yet as I began to prepare the mixing of the fuel.

This was always carried out in the outdoors for obvious reasons! Also, it was our safety policy never to transport mixed chemicals to the launch site. I must add that no smoking was ever allowed within 100 m of the pad during this activity and the range safety officer saw to this stringently! During this time, a car drew up at the bunker and three men emerged. Perfect strangers to all.

A voice boomed over the PA system, summoning me to the bunker! I removed the head-piece of the asbestos suit and waddled over. The two men were from the Chief Inspector of Explosives' office and they were here to see and assess my operation. So after years of the sword of Damocles hanging over my head, it had finally fallen! In many ways I was glad of their coming to witness this shoot, as it was the biggest one to date. If it satisfied them, then perhaps the rope that they had around me would loosen a little. However, they were present and full of questions. I asked to be excused as I had a half-fuelled rocket waiting and I would consider it a personal favour if I could finish. They agreed, but wanted to see what was happening. I could not refuse but pointed out that they would then be in violation of the written directive I was bound to from their boss! For was I not the only one, in an asbestos suit, allowed at the rocket during fuelling? The man who appeared to be the senior of the two seemed undecided, then said that I was quite right and left it at that.

Once the rocket was fuelled, the crew quickly and efficiently loaded it into the tower, sliding the two stages together and checked the launch angle. I got the OK from the crew and now I was ready to answer all that came my way! In the end the government officials appeared satisfied with my explanations to all the queries, but I had a nagging thought that all was not quite what it seemed. The countdown had started, as we were loath to hold a fully fuelled rocket for too long.

I invited the senior man into the bunker, as we were a little pushed for space and he appeared quite impressed with the array of equipment in front of the firing team. He had taken note of the change of coloured flags as each step was completed. The red one was fluttering slowly now. Spectators were moved a kilometre away from the pad when the red flag appeared and out came the telephoto lenses. My two friends from the press from yesterday were there setting up their camera. Jack had the best place of the lot. He had set an 8 mm movie camera on a tripod, just outside the bunker, and he leaned out of the side with a still camera to capture the take-off.

The flight was in charge of the PA system and his voice suddenly boomed out:

'T minus one minute and counting.' This scared the heck out of all of us!

Silence could be felt, as even the bird-songs seemed to stop. Only the beetle noise continued.

'Master?'
'On.'
'Power?'
'On.'
'Radio?'
'On.'
'Heater?'
'On.'

I inserted the safety key into the panel. The milliammeter showed that the smoke circuit had charged and was at the ready.

'... 4, 3, 2, 1. Ignition!'

All the different coloured lights went out on the panel as the firing officer pressed the final button. Immediately 12 V rushed along the 100 m of cable, closing the relay contacts at the junction box. As they closed, 90 V were applied to the igniter in the first stage and at the same time the external batteries were disconnected to the radio and the smoke circuits in the rocket.

Our olooping giant awoko and roared, with fire and smoke, into life. The rocket blasted the ground with a mighty fist of flame, cloaked in wreaths of thick oily smoke. It streaked into the air, the thick smoke trailing in its wake. The ascent was traced onto the blue of the sky by the now thin exhaust stream. It carried on upward at a fast clip and as usual I was counting the seconds to staging. This was when the second stage ignited and separated from the first. It never happened! The vehicle slowed and finally stopped. It hovered for a split second then turned, nose down, and commenced the long fall back to earth. A white cloud marked the blue of the morning sky where the rocket had turned over, a solitary, small white cloud of man-made smoke.

As the rocket lifted off, I had stepped to the side of the bunker so that I had a clear view of the flight. Looking out at the rocket through the pipe at ground level I could only see the vehicle rise about 10 m, so to see the entire flight, I had to look from outside the bunker. My heart sank as I counted past the moment when the second-stage ignition should have occurred. The sequencer for the igniter had once again failed. At that second I swore that never again would I see that type of equipment onboard a rocket of mine!

The impact of the entire rocket was an anticlimax to hours of work by the group. Although one could put a price on the cost of the rocket, what price could one put on the time and energy spent over weeks of endeavours? There was no price! I felt the disappointment of my colleagues as a physical thing. The second stage was buried deep in the ground while the first stage lay some 3 m from it. The time period for a misfire was being observed and nobody was to move away from his position. I unfolded the asbestos suit and prepared to render the loaded stage safe. I handed the official my copy of the safety procedures and left the bunker.

The safety key was in my pocket and I approached the impact point with caution. I could see the igniter wires jutting out of the nozzle of the second stage, and on closer inspection, the igniter appeared intact. Inside the suit I was sweating profusely but the discomfort was nothing compared to the sick feeling that I felt in my stomach. After so many flights, successful or not, the inspectors had to arrive on this day and witness our biggest rocket fail! I felt cheated. I connected the emergency coil of cable to the igniter and gently walked backwards, uncoiling it as I went. I had lifted the 90 V battery from the relay box at the pad on my way to the impact site and placed it on the ground. When I reached it on the way back I lay on the ground and held up my right arm. This was the signal that I was ready to fire the embedded motor.

The loud, metallic voice of the PA system blared out:

'Take cover! X minus 20 seconds.' ('X' denoted a possible explosion.)

I was 100 m from the rocket and I hugged the ground, making myself as small as possible.

'... 3, 2, 1.' I waited a full second and pushed the plug into the battery. A column of white smoke 100 m long shot into the air accompanied by an angry flash and sustained roar! The motor had ignited and not exploded! I felt a heavy weight lift from my shoulders as I stood up and walked to where the last remnants of wispy smoke drifted upwards. Perhaps this was the better way for the two officials to witness what we were attempting. To see for themselves our operation under the worst conditions. The second stage had not ignited and fallen to the ground fully fuelled. There had been no explosion on impact and the safety measures in force ensured no injury or damage to property. The rocket had been rendered safe and the post-mortem would begin.

As suspected, the igniter mechanism in the upper stage had failed to operate owing to the pullout plug not releasing at lift-off. The senior official drew me aside and told me that they had seen enough and were leaving. He was satisfied in the manner that the launch and recovery had proceeded, but how high had the first stage ascended? I showed him the calculations from the tracking crew and they estimated that it had risen to a little more that 850 m. He wrote this information in a book, shook hands with me and departed. My first encounter with the inspectors from the Department of Explosives at an actual launch was over! Whatever the outcome of their report to their superior, it would certainly have an effect on my future research.

Stop or carry on? The answer would not be long in forthcoming.

The various crews packed up their equipment and stowed it in cars and vans and slowly drove away. The rocket crew were the last to finish, as we had to recover the fallen pieces of the wreckage. Digging the deeply buried second stage out of the ground took some time and effort because of the depth it had penetrated into the clinging soil. Coupled with this was the bent and twisted form it had assumed from the impact. At last it was freed and packed into my car along with the rest of the rocket and control panel. I still had coffee left in my thermos and David, Gordon, Jack and myself sat and sipped the steaming liquid. We spoke about the system used to fire the upper stage and all had the same sentiment as I did. Get rid of it!

Ostensibly this sounded simple but in practice it had been the most troublesome, yet complicated device in all of the multi-stage rockets we had fired. That was it! It was too delicate and the electronics were not simple. We had seen if the action of the plug on the side of the rocket was not pulled out completely when the lift-off occurred, the timer inside failed. A far more simple and reliable method had to be sought. And this was the hard practical aspect. Simple? Wrong again! Nothing in a high-power rocket was 'simple'.

A number of ideas were thrown around on that sunny Sunday morning. The smell of the exhaust hung in the air, as if taunting us to find the solution. We sat there for the most part of the morning and our brainstorming yielded a great big blob! Nothing. There had to be a way out of this that would ensure a safe, positive ignition every time we fired a multi-stage job! But what? A mechanical device? No, too temperamental. We had gone the electronic way to be more cost-effective and reliable only to find that the cost was justified but the reliability edged around the 50% mark. In a rocket that was a one shot-off event, 50% meant that half of the vehicles would perform and the other half would fail. What was it that Pascal had said about experimenting? He was so right – it brought home what we were learning on the launch pad.

With every rocket I had learnt a little more. Whether it worked or failed, something new went into my head. Driving home that Sunday was a revelation to me. Stop the big boys and go back to designing smaller, cheaper multi-stage rockets that worked every time the button was pushed, and once that was achieved – go for the big fellow!

Large boosters were essential for the first stage in multi-stage rockets. It was this stage that lifted the vehicle from the earth, straining to overcome gravity and giving the upper stages the necessary velocity that was unobtainable with a single-stage configuration. To fire a large, single-stage rocket had been accomplished by the group but we had suffered singularly bad luck with multi-stage stacks. Each time a failure was recorded it was because of the ignition sequencing for the upper stages. These devices had been a bone in my throat for a long time but the answer to a simpler, reliable system kept eluding me. My thinking was wrong! I had selected a track and relentlessly followed it, believing that the solution was to be found along it.

How wrong could I be? It was the very nature of science that anything that had a possible 'bug' in it did not become more reliable as time went by, but either got worse or stayed the same. I had chosen the wrong path by sticking to the thing that worked 50% of the time. This percentage was 'not on', for no sane person would take a 50% chance on his or her life! The odds must be a heck of a lot better. This thought seemed to clear some of the cobwebs out of my thinking and more rational, although completely different, schemes began to germinate in my head. For one thing, the rocket produced so much heat at ignition that it engulfed the vehicle before it left the pad. This extremely high temperature was capable of destroying anything that was remotely exposed and second, the tremendous acceleration at which the rocket left the tower also complimented the enormous stress that was imposed on all the systems inside the instrument compartment. Looking back, it was pure luck that the first two-stage rocket functioned as planned and exceeded the 1,6 km mark.

I must mention that after the successful flight of the two-stage rocket that cleared 1,6 km, I had written a long letter to Dr Wernher von Braun, the well-known rocket scientist at Redstone Arsenal in Huntsville, Alabama. It was he that was responsible for the launching of the West's first satellite, carried into orbit by his modified Redstone rocket in January 1958. In my letter to him I described exactly what my goal was, as well as the firings we had exacted. After months of waiting, a reply to my epistle was received! He was indeed interested in the projects that were being undertaken in this country and wished me the very best of luck for all future attempts. I felt and still feel a great respect for that man.

It was his giant Saturn V rocket that saw man land on the moon in July 1969 and return safely to Earth. That was 32 years ago and to this day his name lives on. His name and the success of rockets as viable space transport systems are synonymous. Just to gaze upon the huge F1 engine that he was responsible for is enough to say: 'Here is the master.' Five of these engines clustered together lifted the Saturn V off the pad and sent it into space, developing more than three million kilograms of thrust!

Herein lies the answer to the multi-stage rocket. It would have been inappropriate of me to ask him to solve my petty problem, for whom could he turn to with his frightful responsibilities? No, I would indeed have to find the right system myself. Why was I steeped in blankness when it came to a problem that had been looking over my shoulder for so long? I had to take a different tack on it now.

My memory flew back to October 1957 as I stood in the front garden of my home, tears streaming down my cheeks as I stared into the dark night sky. A pale star was moving at speed across the heavens and signified all that I was trying to achieve. My emotions were whipped into the darkness as I watched Sputnik, the first satellite launched into space. It did not matter to me that the Russians had succeeded whilst the USA were still struggling to get a launch vehicle off the pad in one piece. What mattered to me was that it could be accomplished. And my tears that night were tears of happiness! The dream that I had so long ago was now flashing across the sky, announcing to the world that a rocket had placed this 'thing' into outer space. All I had ever asked was to be given the opportunity to attempt the same, and given the facilities and backing, I was sure that it was possible.

Undoubtedly the USA had problems with their hardware, but equally they would surmount them and also take their place in the vastness of space. They did not have long to wait! Dr von Braun's Explorer 1 was only four months away! The elation the Americans felt could only be imagined as their satellite swung around the blue planet, transmitting its radio signals back to the scientists.

When all aspects were considered of a sequence ignition system with the equipment that was available in those days, the prospects were rather bleak! It had to be able to withstand the g forces at take-off as well as be impervious to the hellish flash of ignition. The solution was not in thermionic valves, as solid state technology was not yet available, and transistors were just about to make an expensive and dramatic leap into the spotlight. So another route for my dilemma was being signalled. If this one could be negotiated, the very sky stood to be cracked wide open.

When the final solution presented itself into my fogged brain, it was the very thing I had wanted no part of! A mechanical method! So simple in operation and not at all difficult to construct. Would it work every time? Experiments under all sorts of conditions were mapped out. I was timorous in calling a meeting of the rocket crew to explain my rush of blood!

They studied my proposal in silence with only the sound of pencils on paper as they copied the drawing on the table. At length the questions began, and answers on my part were cautious and as accurate as I could give. It was my belief that we, as a unit, were sceptical as to the functioning of a mechanical solution. It looked good on paper and was ostensibly not difficult to produce. But how would it perform when subjected to the adverse conditions previously described?

The only way to get the feedback was to apply it practically in various vehicles, starting with small two-stage rockets and, if successful, moving onto bigger and more powerful versions. The various members with appropriate machines and skills did not hesitate to offer their services and a working drawing was produced. This new unit would be fitted into a known two-stage rocket that had failed. (The failure was caused by the second-stage igniter not being activated by the suspect sequencer and not any design fault.)

Another rocket was built exactly according to the one that had misfired. It was identical in every aspect except for the forward bulkhead. Here lay our latest innovation! All our aspirations were riding on this modified piece of hardware and needless to say, the launch date was set for the earliest possible day. If it worked, the rocket would reach 2,4 km. Not an earth-shattering event as far as the altitude was concerned, but a step along the right road this time. If the unit proved sound, it could even be salvaged and utilised in another rocket of the same diameter.

Optimism ran high and instead of building one rocket, three were constructed along the same lines. The launch date was pushed back to enable the three to be fired on the same day. It was certainly ambitious and we realised the disappointment that awaited us if this method of sequencing failed to live up to our expectations. We had ventured onto the launch site, how many times, with high hopes only to see our creations work every second time. It was proving a waste of money and detrimental to our nervous systems! As I said in the beginning: 'Who said it was going to be easy?'

Chapter 6

Mighty flights by courtesy of Prometheus

The day finally arrived when the three experimental rockets would be fired and I could feel the pent-up excitement of the crew as they set about their preparations to ready the tracking and erect the tower. They went about their tasks in silence. The usual banter was missing and in its place was an air of solemnity. This added to the feeling of expectancy and did nothing to allay my own emotions. Over the years I had learnt to deal with failure. I did not like it but had to accept it as part of the unusual pastime that had become my life. I tried to analyse my feelings and it did not require the services of a psychiatrist to hit the cause on the head!

The group, as a whole, looked to me for the success of each launch and expressed their utter joy when things worked as planned. If things did not work out, I could feel their disappointment as a tangible thing. I had learnt to read each of their faces, happiness or hidden misery after each launch. They were my friends and helpers, without whom I could never have got so far in my quest to build better and bigger rockets and achieve my ultimate goal. What had started as an attempt to make a small spaceship that would take to the air belching smoke and fire had now become a supreme effort to place a large rocket into space.

I had understood from the moment that my first 'big one' spluttered into the air all those years ago that I would seek the ultimate. A satellite launch vehicle! Many obstacles had come my way and, no doubt, many more were on their way. But to give up because of complexity or adversity was not on. Too much water, among other things, had flowed under the bridge even to contemplate such a step. So it was with a sense of duty to my colleagues that I felt honour-bound to make every Saxon launch an event to remember. Hence the butterflies! Once I realised this, I felt better knowing that, whatever the outcome, they were in it with me.

I had completed fuelling the first rocket and was busy assembling the two stages together when I glanced back at the bunker. It seemed strange not to see people gathered at the parking area but I had not told anyone about these experimental shots. This was purely and simply only for us. The rocket slid easily into the tower and all that remained was for me to connect the first-stage ignition leads. The usual tracking system was located in the section just aft of the nose and the charging cable to the panel connections had been completed.

The yellow flag was being replaced with the red one as I entered the bunker. The smell of the soil, the quiet voices, and the row of equipment was so familiar to me that I immediately felt at ease. Gordon, as always, was the flight safety officer that day and I asked him if the time to launch was as planned. I got the thumbs up from him so I took my position behind the firing panel. David looked around at me and nodded his approval as to the readiness of the firing panel. The various indicator lamps cast an eerie glow in the dimness. All OK. Now the wait as the clocks were checked as well as receiving the go-ahead from the far-away trackers. All was now ready and in a few minutes the firing would be out of our hands and into the lap of the gods.

Rocket #1 stood tall and white on the pad as the countdown started. Last-minute checks were carried out to make sure that whatever had to be done had been done.

'T minus 10 seconds.' Gordon's voice sounded loud in the confines of the bunker. I had been watching the charging circuit meter, the indicator that the capacitor was ready to ignite the smoke charge edged into the red. It was fully charged with 360 V. The entire panel lights winked into life. Rocket clear!

'… 3, 2, 1. Ignition!'

Now the sight that was familiar to the group! The first stage lit up with clouds of white smoke blasting out of the exhaust duct accompanied by a brilliant yellow and orange flame lancing out of the tail. Lift-off! The slender shape streaked into the air, gathering speed as it ascended. Suddenly a second blast struck our ears, but spontaneous cheering in the bunker drowned it out. The sequencer had worked! I stepped out of the bunker and stared at the rapidly diminishing white spot against an azure blue sky. My heart hammered in my chest, as the small cloud appeared high above my head. I listened to the whistling of the descent for some time before it abruptly ceased. The rocket had impacted. All clear, so up the green!

Tracking came in with their figures. The altitude reached was 2,3 km! Not bad for a first-off sequencer! The recovery crew sped off to retrieve the fallen two stages for examination.

In the meantime, preparations were under way to launch the next rocket. This one was smaller than the previous launch and was designed to reach 1,6 km. The crew arrived with the two pieces of the blackened rocket, and we wasted no time in disassembling it. The bulkhead of the first stage was removed and our mechanical device was adjudged fit for another launch. Now do not get too excited, Desmond, this was only the first shot. Still two more to go before any conclusions could be drawn.

The second rocket was readied and erected on the pad. It was painted all white with no black lettering. The procedure to launch progressed without any problems. The ignition command was given and the lift-off, although less noisy than the first, was a scene straight from science fiction! The smaller rocket whooshed out of the tower and headed vertically away. Just as the smoke and flame from the exhaust appeared to thin out, another burst of flame erupted from the upper stage, sending it arrow-like upwards! The second stage had also ignited right on time!

The tiny dot disappeared momentarily as a white cloud suddenly blossomed into an irregular pattern. The first tracking marker was now only a faint smear of mist in the blue as it was torn apart by the phantom winds on high. The dot began to gain in size as it approached the ground, residual exhaust streaming behind it. It impacted with a muted thud into the ground. Inspection of this rocket's first-stage bulkhead revealed the same results as the first ... no damage! The score was now two out of two! Was the nemesis finally laid to rest?

The third rocket was actually a single-stage job with a pyrotechnic package instead of a second stage. If the pyro ignited, it would demonstrate that sufficient energy had been released to ignite an upper stage. This rocket was powered by one of our largest motors and if our sequence device worked on this motor, the doors would be opened wide! The acid test was about to start. As we retired to the bunker for the third countdown, the firing team exuded a feeling of confidence.

No tracking smoke charge was installed, as the pyro would act as a marker and also as a retro fire to significantly stop the rocket's upward journey. I was not a superstitious person by nature, but I had an overwhelming desire to cross my fingers or do something equally silly as the bottom of the launch tower became obliterated by the angry expulsion of the motor start. Dust and small stones were hurled into the air as the flame-filled smoke beat at the ground. The loud bang of the motor filled the bunker and as always, we were left speechless as we witnessed the fiery lift-off.

The tall 1,8 m spire seemed to be shoved skyward by the thick smoke column racing from the tail. A slight wind, coming from the south, wafted some smoke into the bunker, causing an assortment of coughs from us.

'Heater off. Power off. Master off.' Even flight's commands were interspersed with bouts of coughing!

I took up my usual position outside to view the rocket surging into the air. A dazzling display of red and green sparks lit up a small area of the sky. The pyro had ignited! I yelled as loud as I could at the success of the third and final experiment. We had got ourselves a system to work each time and the future looked most promising! With the advent of the firework display high above the Transvaal veld, the now-silent rocket had began its long fall back to earth. As it gathered speed, the wind passing over the guidance fins caused a loud whistling sound, which gained in intensity as it neared the ground. As the rocket struck, the sound immediately ceased, leaving the vastness of the firing range in a deafening silence.

The remains of this vehicle joined those of its predecessors on the ground alongside the bunker. All three of the ignition systems were removed perhaps to work another day! Although blackened by heat, they were undamaged and were a revelation to us all. A really uncomplicated mechanical piece of equipment had solved our seemingly darkest problem! It was a permanent solution and would see our multi-stage rockets work flawlessly in the future.

In retrospect, as I look back on those exciting times, I am awe-struck that the answer to our work/do not work system had not been seen earlier. It is a true saying that the woods cannot be seen for the trees. When you are immersed in a problem there is a tendency to try and rectify the malfunction with quick-fix solutions and it is my experience that this line of thought never works in the end. You may have partial success as we did, but finally you have to say: 'Fine, that did not go so well so let us try another method. The new method may be just what we are looking for.'

This was true of my situation. The thing had gone according to plan on that one occasion, but had failed this time. It will be used again and maybe it will work this time round. Wrong! I have never known a questionable item to be reliable. Another lesson had been learnt and I was the wiser for it.

The answer: when the rocket motor is ignited, immediately there is a tremendous rise in pressure and temperature as the fuel releases energy. (The starting transient.) Depending on the propellant grain configuration, the combustion chamber pressure may increase or decrease as burning time elapses. The temperature rises as burning time progresses. With a little thought, these aspects may be used to the designer's advantage. I latched onto the pressure to bring about a workable sequence ignition system!

By utilising a 15 kg compression spring and attaching it to a tempered steel plate over a hole in the forward bulkhead, I had, in the simplest terms, a spring-loaded one-way valve. When the motor was ignited, the high-pressure gas in the chamber slammed the plate closed as the internal pressure force on the plate was greater than that of the spring. At the end of the burn the pressure in the chamber became less until the force on the plate fell below 15 kg, at which stage the spring forced the plate away from the hole in the bulkhead. This opened up the hole to allow the last of the high-pressure high-temperature gas to flow out of the bulkhead and into the nozzle of the upper stage, effectively igniting the fuel in that stage.

In fact it was so reliable that with this device a *five*-stage rocket was built and successfully fired! Years later when solid state technology was used in household items, firing systems became a walk in the park, but in the early days, they caused headaches for all multi-stage rocket designers. So much for breaking one's head over problems when the answer was about to bite you! The feeling at the range when the three shots were completed and the data collated was one of relief. I, for one, felt drained yet had also a sense of accomplishment. The three different configurations had worked as designed, all using the same delayed ignition system. I had taken a new road and the results were more than I had ever hoped for. It was now time to go for the really big 'jobs'.

I felt that the large two-stage rocket that had failed when the three officials were present was a good vehicle to utilise our new system to ignite the upper stage. I unrolled the drawings of it and pored over them. A few minor changes, such as a new fin planform and slightly longer equipment housing for the upper stage were the only changes. It looked almost the same as its predecessor and I decided to paint the first stage in our usual white and black, but used yellow for the second stage. This gave the rocket a character of its own! The same method of coupling the stages together was employed, as it had always been the simplest and most effective. The completed vehicle stood 3,35 m tall and weighed just on 70 kg. It was the heaviest rocket I had designed and my calculations indicated that an altitude of 3,5 km could be reached.

The day of the launch of Saxon Alpha 19 (It was my 527th rocket!) dawned with a chilly wind and white frost on the ground. The rocket crew had braved the cold night and slept in the bunker where we were protected from the wind. The two sections of the rocket were wrapped in canvas inside one of the vans, as I certainly did not want a wet thing on my hands in the morning! Gradually the talking in the bunker had dwindled as sleep took over and most of my friends succumbed to it. I lay in the darkness with many thoughts running through my mind until I too fell asleep.

I awoke to the sound of voices and movement as the crew came to life. I stood up and went outside into the cold light of day. I felt awful and hastily went to my car and washed my face and cleaned my teeth. I felt a hand on my shoulder and a steaming mug of black coffee was held in front of my face. Here was the elixir to revive one. I thanked the kind soul (it was Gordon) for saving me and gratefully sipped it. Life was returning to my body and within a few moments, I came fully awake. The sun was beginning to spread its welcome warmth over one of its children.

In the dim light the final tasks were attended to and once the sun was fully above the horizon, I began preparing the propellant. Fuelling was slow and time consuming but eventually it was completed. In the beginning I was happy to be in the asbestos suit as it protected me from the chill air, but after some minutes it was a different story! The sweat began and I had a job keeping it out of my eyes. I signalled to notify the crew that the rocket was ready to be loaded onto the pad and waited for their arrival. It took three men to slide it up the tower and settle the tail and fins over the exhaust duct at the base of the pad.

Back at the bunker I swapped the suit for a woollen jersey and joined the firing crew at the panel. The yellow flag had been hoisted so the range safety officer saw to it that nobody went within 100 m of the rocket. All electrical connections were made and the various circuits tested. All OK. The flight checked and tested the PA system and his voice boomed across the half-awake range. Birds that had been unseen by us took to the air in panic at the sound of this sudden blast of sound. The peace of the veld would soon be shattered by a sound much louder than this!

All systems on board the rocket were checked out and signed off by me. I had told the group that it was more than likely that the government people would show up and that if we were finished ahead of time with our readying the rocket, we would wait until the very last moment before firing. I had informed the inspector well in advance of the proposed launch date and as usual I had appended his written permission to my clipboard.

See page 113 for enlarged version

The clock ticked on and the government men did not arrive, so I gave the word to go ahead. The gentlemen of the press were notified that the count was about to begin and they readied their cameras. The cables were connected to the igniter and the smoke charge. The transmitter in the rocket was switched on and checked. The tracking men had a long walk and we waited for them to signal their readiness. On receipt of this, the countdown commenced. Back in the bunker, the whine of the transmitter in the rocket seemed unusually loud in the silence that always preceded the count. I checked the signal strength meter and the needle was holding steady. The high voltage-charging meter indicated that the capacitor was fully charged.

See page 100 for enlarged version

'T minus 15 seconds and counting.' Again flight's voice scared the countryside.

I waited a few seconds and inserted the safety key into the panel. The count proceeded as normal:

'Heater?'
'On.'
'Radio?'
'On.'

See page 103
for enlarged version

'Rocket clear.'

'...3, 2, 1. Ignition.'

This lift-off was spectacular! The ground shuddered as the rocket blasted its way into the air; the thick, long pillar of smoke, interspersed with flame, stretched like a giant serpent into the sky. A thunderous bang beat down from the heavens as the second stage ignited, only to become muted as the missile disappeared from view. The needle of the signal strength meter dropped slowly and had to be kept under close surveillance to catch the lowest point it reached. I was ready with a wax crayon to mark the glass in front of the meter. The needle began to rise and I hastily marked the glass.

I stepped outside the bunker and was greeted by the high-pitched whistle of the descending rocket. I stood and stared at the still-visible vapour trail scribed on the blue of the sky. It had the appearance of a tightly bent horseshoe – ascent, turnover and rapid dive earthwards! It must have been very cold up there! The impact was seen but not heard. The second stage had landed due south of the pad, exactly as planned. We would have to look for the lower stage, as I was sure that all eyes were on the slowly dissipating, man-made cloud high above the bunker. The whine from the on-board transmitter had ceased some seconds ago as the impact occurred. I watched the remnants of the exhaust slowly snake upwards, then went back into the bunker and witnessed the equipment being methodically switched off. I stood and wrote down the altitude as indicated by radio but waited for the tracking crew to arrive with their results.

They were all smiles as they joined the rest of us, now all at the bunker.

'Three thousand six hundred metres! We have checked again and again and get the same answer each time. You will want to check it yourself but hell, that was a great shoot!'

'No need to, it is so close to the radio info. You fellas are spot on! The signal from the rocket gave a height of 3,5 km! So how is that?' Smiles all round and why not? We had come back from a misfire with a rocket that size and succeeded in firing it! I got into the van that one of the tracking men used and drove out to recover the two pieces. We chatted on the way and first collected the blackened lower stage. (The black colour was actually from the second stage igniting.) I peered into the forward bulkhead and the sequencer appeared to be in one piece. (In fact, it was ready for another shot when we examined it.) The second stage displayed only its large fins sticking above the ground, the short grass gently swaying in the light breeze.

To try and pull it out was impossible even with three of us really putting our backs into it. The answer was to tie a rope to it and to the van, slowly backing away. The embedded stage gradually emerged, bent into a 'Z' shape. It had obviously struck a buried boulder on its downward plunge and this was the result. No wonder we could not budge it! The wreckage was loaded onto the van and we drove back to the bunker.

As we approached the launch area, I was amazed to see the effect the lift-off had on the ground! A circle of about 15 m in diameter was a light grey and around this circle was another. The outer one was blackened and burnt, extending for a good 50 m beyond the grey inner circle. This was indicative of the exhaust effects as the rocket rose into the air. As the rocket went higher, the angle of the fiery jet blast from the nozzle increased, scorching the earth in a giant circle. No small stones were to be found within the boundary of this circle; all loose items had been blown violently away. I had witnessed this scene before but never on such a scale.

Other members of the group had efficiently dealt with small grass fires that had been ignited by the rocket blast. Fortunately, this was the only untoward happening from the fiery lift-off and I was very thankful that this range had been offered to us. The short shrub grass that grew in that area was by no means a latent threat for a serious veld fire and also the tower was erected on a flat, sandy patch. No fire extinguishers were needed, as simple beating was sufficient to smother the flames that dotted the discoloured blast area.

The first stage was dismantled to inspect the ignition device in the forward bulkhead and, although blackened, it was found to still be in working order. The pieces of the rocket were loaded onto a motor vehicle and the launch site cleaned and left as we had found it. Someone produced more coffee and we sat in silence, with pleasant thoughts of our own. We had succeeded with the largest rocket to date and Jack had managed to photograph the rocket about 20 m above the pad, the familiar smoke column pushing it upwards. The report to the inspector of explosives was going to make very pleasant reading and I looked forward to handing it to him!

Chapter 7

The final countdowns begin

The year 1962 was a great one for the SARRG. It had conducted four highly successful static test runs using a new fuel that offered increased performance over the zinc/sulphur mixture that had been used in most of our previous motors. This new propellant was a combination of asphalt and potassium perchlorate. If our old motors exuded copious amounts of smoke, this new mix was the grandfather of all smokers! An advantage was its slower burning rate, which meant that a guidance system I had been developing could be put to work.

Up to the last launch, all the rockets had depended on the four fins at the rear to give them arrow-like stability. This was achieved as long as the centre of aerodynamic pressure was behind the centre of gravity of the rocket. This would hold true as long as the vehicle was operating in the atmosphere, but with no air to flow over the fins at extreme altitudes the controlling moment of the fins would be lost.

With a longer burning motor many options were available. First and foremost, thrust vector control was possible! This means that by deflecting the exhaust flame coming out of the nozzle at an angle to that of the flight axis, the rocket can be steered to alter its original course. Thrust vector control could also be used to correct for the effects of strong gusts of wind as the rocket left the tower. A gyro sensor pack could detect the deviation and the control system could then correct for it. This means of control can be used at high altitudes and at low velocities when the fins are not effective.

For a satellite launch vehicle it is imperative to gradually alter the flight path to one that is parallel to the Earth's surface before the final-stage motor burns out. This gives the vehicle the necessary velocity to achieve earth orbit at the desired height above the earth. With modern-day launch vehicles the necessary alteration of trajectory is accomplished by mounting the liquid fuel combustion chamber(s) on gimbals. This allows the motor to be swivelled in any direction, either to stabilise the rocket or alter the flight path. Again, any deviation is sensed by the gyro and a command is transmitted to the servo actuators to swivel the motor to change the thrust axis.

Let it be assumed that a vertically ascending rocket enters a high velocity windstream and the nose begins to point to the right. This deviation is sensed by the gyro and it sends a command to the actuators. To correct the swing to the right, the nozzle will swivel to the left, imparting a side thrust to the right side of the tail. This counteracts the right movement of the nose and returns the rocket to its prescribed course. But as the rocket begins to move back, the gyro sends another message to swing the nozzle to the right to damp the initial correction. If left undamped, the nose would swing too far to the left and another large control moment would be sent, this time returning the rocket to the right again and back to the left and so on. The rocket would snake all over the sky until it destroyed itself!

I knew of these systems and with a longer burning motor they could be applied. Hence the desire to build a vehicle that would take South Africa into space. My first attempt at thrust vector control was under way on the drawing board. It consisted of a sheet metal cylinder 15 cm long and 10 cm in diameter. This was loosely fitted at the rear of the rocket nozzle and was linked to the instrument housing behind the nose by four 4 mm rods. Two opposite rods were attached to a bell-crank so that if the right side went down, the left went up. The same applied to the other two rods. A 5 cm gyro was fitted in the nose with four electrical contacts on its cage. These were insulated from each other and were in close proximity to a carbon disc. Voltage from a battery was applied to the contacts but the air-gap between them and the carbon prevented any current flow.

Two small electric motors were coupled to the bell-cranks via a gear train to slow down the movement of the crank. In operation, the gyro would always remain spinning at the angle it was set. If the rocket deviated from that angle the actual rocket would move, but the gyro would remain at the pre-set position. This movement would cause the electrical contact to touch the carbon disc, allowing voltage to flow to the motor that would in turn push the bell-crank. This action would move the appropriate rod attached to the cylinder at the tail, deflecting the exhaust. To deflect in the opposite direction, the carbon disc was in two semicircles, one side positive and the other negative. By changing polarity, the required rod was operated. Full deflection gave 3° to 5° of off-set. On reaching the end of its travel, the rod bumped a reverse switch that returned the bell-crank to neutral. Great fun!

A 1,8 m rocket was built with this control installed and we waited for a really windy day to hastily launch it. The expected altitude was 1 000 m and we would be able to see, at least partway, whether the thrust vector worked. The waiting was a blessing, because modifications to the pad had to be undertaken to accept the new tail configuration and this took time.

I must take this opportunity to relate the meeting I had with the inspector after the last successful two-stage launch. If I expected a pat on the back, I was horribly mistaken!

'How high is high enough for you and your bunch?' was the response after the report and photographs had been well and truly studied. I sat dead still, not knowing what to say. Over the many years I thought I had got to understand this man, but his remark left me speechless. He peered at me over his spectacles, a long-suffering look on his face as he awaited my reply.

'Sir, you know more than anyone else, after all this time, where my goal lies. We are making steady progress in the development of a satellite vehicle. You have seen the reports from day one and surely you are now aware of our objective.'

'Your "objective" is a pipe dream and nothing more. These things cost a great deal of money and where do you think you will raise the funding?' His quick retort was brusque and sarcastic. He knew about my sponsors and he also knew that I knew that the project would cost a bomb!

'Surely the government would gain prestige from being one of the nations that are capable of having their own satellite? Are not all nations interested in the quest for space exploration? Sir, I started this little venture when I was a child and today I am a husband and father. In the years that have passed, have I not complied with your every directive to the letter? Have I not invited you or members of your staff to attend every launch only to be visited once?' I was getting hot under the collar.

'You, my boy, are stretching for things out of reach and I cannot see your aspirations remotely bearing fruit, even if you had the backing! This much I have decided, if you shoot a rocket in future and it is expected to go a kilometre, the range must be a kilometre in radius. So, whatever the expected altitude, that must be the radius of the piece of ground.' I found this statement to be absurd! He either had no idea of ballistics or was just being bloody-minded!

'It is a known ballistic fact that a rocket will follow the same angle down as it had going up. In my letters for permission to launch, I have given you the predicted point of impact and each time the rocket has landed within those figures. To say that the radius of the ground must be that of the expected height is not realistic! You are saying that if I launched a rocket to one hundred kilometres, it would have to be fired on a tract of ground two hundred kilometres in diameter, with nothing on it! No such piece of ground exists in South Africa. I have demonstrated that I have control of a rocket in the air, ensuring a safe flight with no chance of causing damage or injury yet you are now making things almost impossible for me!'

By now I was trembling with frustration. The injustice after all these years! I saw now that no real interest had been there right from the beginning. I had been allowed to play with my toys but now playtime was over. Why? Most of the technology had been understood and put into practice but with the goalposts in front of us, the game had been called off. I was given the conditions and woe betide me if I did not follow! I was curtly waved out of the office and dismissed. To stay and argue was indeed pointless.

I left that office with sad thoughts in my head and completely deflated. I felt as if I had been misled right from the beginning. The realisation that I had been given a short piece of rope to play with angered me. The anger was because I had been under the impression for so long that with each success I had gained more credibility with the authorities. But now I saw that my efforts were so much an inconvenience in time and paper to them. Perhaps they had thought that with the passage of time my enthusiasm would ultimately have died. As this was not so, the gloves had to come off and conditions that were unreasonable now came my way.

The chief inspector knew that I would not be satisfied with using rockets as a Sunday toy to be taken out and popped a few metres into the air.

He had seen the development grow over the years and possibly thought that this thing was becoming too large for him to handle. If that was true, why had I not been told that enough was enough and that was that!

Instead, clauses were brought in that were clearly meant to bring me to a self-induced stoppage to activity in this field. I felt numb! But at the same time the wheels were turning. What about firing out over the sea? There was an idea! In the meantime our present range was more than adequate for our immediate needs.

My priority was to test the thrust vector control as soon as possible mounted on a rocket that could be observed throughout its flight. The test would be to launch the vehicle vertically and to alter the flight to that of a horizontal trajectory. That would be the cherry on the top! If successful, it meant that we had the means to programme a large rocket into a satellite orbit attitude. With this in our bag of tricks the development of the launch vehicle was going to take a mighty effort! But it could be done!

Before becoming too optimistic, I deemed it necessary to launch a 1,8 m rocket in the most unfavourable windy conditions. This would be to verify the degree of effect that could be imposed by the steering mechanism. Usually when launched during high winds prevailing, the rocket had the inclination to cock into the wind and not be blown in the direction of the wind! If the system performed, it should hold to the original flight path.
The rocket was built and was to be lifted by our standard 300 kg thrust motor that had been our workhorse through many test flights. The anticipated height that it would reach was in the region of 1 000 m and that was high enough to see the results. So with growing impatience, we awaited the day when suitable wind velocities would blow our way. What preparations could be made were made for the up-coming flight and no special equipment was to be hurled aloft in this shoot. I must have become a pain in the rear to the chaps at the Weather Bureau with my daily telephone calls. No, there are no indications of any high winds in your area at present! This was my daily response from the bureau. I left my phone number with them and they promised to call when any change looked imminent. I could hear those esteemed gentlemen saying: 'This chap must be looney-tunes asking for high winds when everyone is happy without them!'

At last the call came; winds gusting up to 25 km/h were expected the next day! The entire group was unable to attend at such short notice but enough men were available to get the launch under way. At the range the dust was blowing fit to bust as the wind whined through the grass and sent a ghostly sound across the wide expanses. Fuelling was a nightmare! It took longer than normal, but at last the rocket was in the tower and connected to the panel. Just looking through the observation pipe showed the short grass and bits of vegetation waving violently about. I had the feeling that nobody in his or her right mind would ever contemplate firing in this weather. To put the lid on the whole affair, light drizzle began to dot the landscape. The sound of drops of water added to the ghostly whistling of the wind.

The panel was brought to life and the count commenced. The loud whistling as he called the time muted the flight's voice.

'... 3, 2, 1. Ignition!' How many times had that word been spoken over the years?

Although it was 08:00, the gloom of the adverse weather gave everything a greyish colour and a feeling of unreality. The motor shot a stream of fire from its exhaust that appeared far brighter than we had seen before. The brilliant flash and clouds of smoke accompanied the loud bang as the rocket left the haven of the tower. Up it sped and at a height of approximately 50 m it appeared to veer to the left, into the wind. Almost as fast as it had begun to change course, it started to return to the vertical! The rest of the flight continued along the prescribed path and was normal and the anticipated height was attained. Jack's photograph of this correction is a rare one indeed!

See page 101 for enlarged version

Apparently the rocket had entered a layer of wind that was at a higher velocity than at ground level and this had the effect that we sought. The rocket had been effectively deviated from its initial course and had returned to the original flightpath. The thrust vector system had worked!

To elaborate on it to meet the requirements of a much-larger rocket was not going to be easy. The problem would be in the procuring of the necessary servomotors as the other bits and pieces were readily obtainable. At about this time, a project was being proposed by the group for a high-altitude rocket that would leave all our other attempts far behind. They were all aware of the restrictions placed on me, but they felt that it could be achieved! A stretch of land would be found even if it had to be fired in the Kalahari Desert! We now had the means to ignite multi-stage vehicles so why wait?

The guys were really fired up and their confidence was infectious. I said I would draw up some preliminary designs to see what they would accept as a 'high altitude' shot. This was the way to go and not sit back and bewail one's fate! Page after page landed in the wastebasket as I discarded one plan after another. A multi-stage job had to be the one to get to the height that was envisaged but not of the proportions that had so far been fired. It was to be a big one!

The final configuration was, to say the least, alarming! It towered 6 m high and comprised five stages, weighing a hefty 105 kg! The final stage was 1,4 m long and carried the transmitter behind the nose housing with which the altitude reached could be ascertained. The pencil-slim rocket with its 20 fins was a sight to behold! It was to be powered by a mixture of zinc dust and sulphur in the lower two stages and a mixture of potassium perchlorate and a polymer in the upper three stages.

It certainly was a breakaway from all previous concepts and the calculated altitude was 25 km! The velocity of the final stage would be in the region of 1 760 km/h, which meant that supersonic speeds would be reached.

The motors had been statically tested and the new propellant worked very well and had a burning rate of 5 cm/s. Compared to the old fuel that had a rate of 229 cm/s, this fuel seemed to burn forever! The launch tower had to be enlarged to accommodate this new vehicle and this was a major undertaking. The rails were not stable at that length of 6 m so a lattice had to be constructed to steady the entire assembly. This included holding the rocket perfectly aligned during the boost phase. All things are possible and within a month most of the alterations had been made, with only the

See page 102 for enlarged version

It was very uncomfortable trying to work in such conditions
Photo: Jack Holloway

The first three-stage rocket in South Africa. My dad is looking over my shoulder as I peruse the pre-flight check-list Photo: Jack Holloway

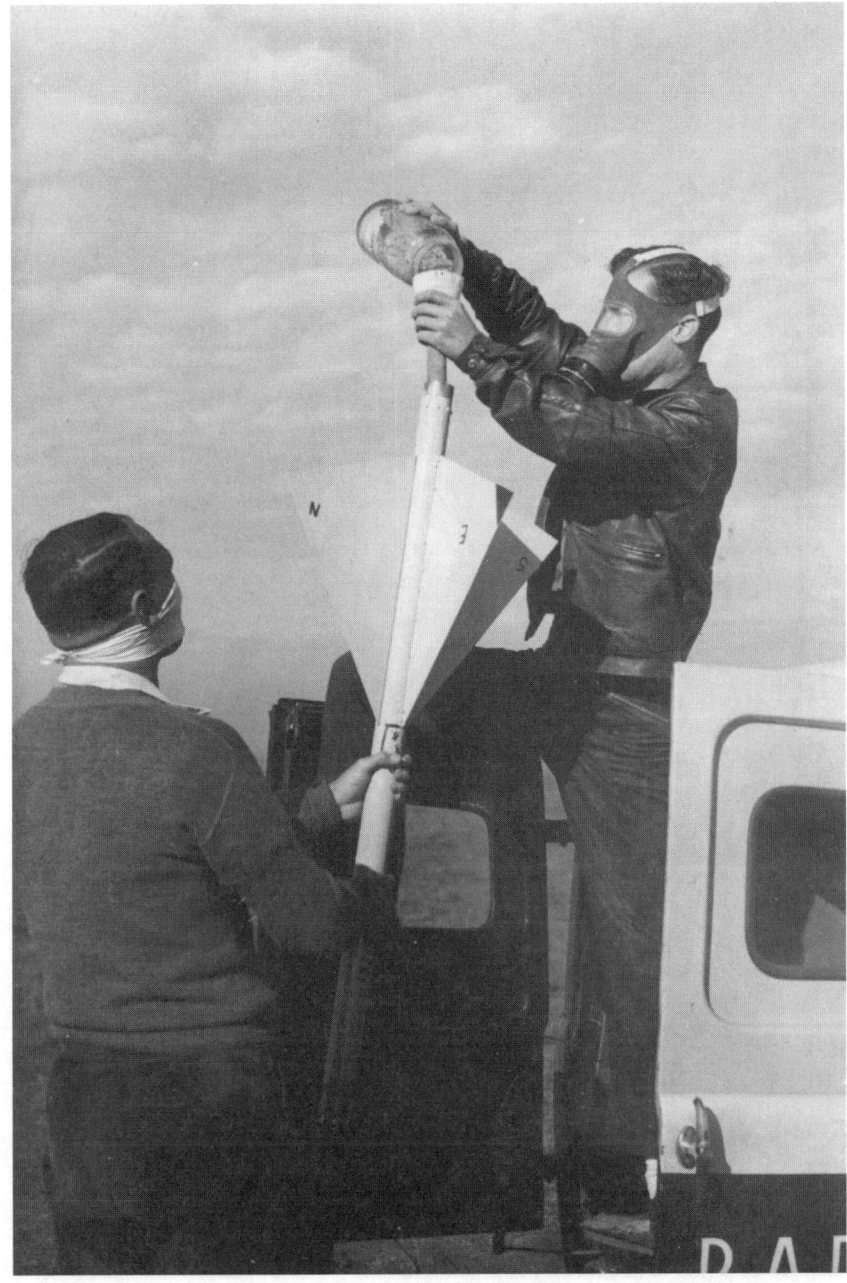

Fuelling a solid fuel motor complete with gas mask
Photo: Jack Holloway

The 100 m smoke column. Take-off, Saxon 18, 19/10/1958; 8 ft 2in single-stage   Photo: Jack Holloway

This is how the cartoonist of the *Rand Daily Mail* saw me after the launch of Sputnik in 1957. Caricature by Bob Connolly

Saxon 18, 19/10/1958. Wiring the 8 ft rocket for firing, 8 ft 2in single-stage. Gordon is the big fellow helping with the job on hand Photo: Jack Holloway

We manage to extricate the bent and twisted final stage from the clinging earth
Photo: Jack Holloway

The firing officer and I enjoy a light moment at the pad
Photo: Jack Holloway

The thrust vector control at work. The path of the rocket can be seen being corrected
Photo: Jack Holloway

The five-stage rocket (not all of it shown) erected in my driveway. The trim tabs on the fins can be clearly seen   Photo: Jack Holloway

The large two-stage rocket 40 m into the air immediately after ignition
Photo: Jack Holloway

Not all rockets went where they were supposed to. This one detonated on ignition
Photo: Jack Holloway

The tower is flattened by the blast of lift-off
Photo: Jack Holloway

Top. We recover the final stage after it came down from 15 mls. Gordon is on the left
Photo: Jack Holloway

Bottom. First of the Saxon series with author Desmond Prout-Jones, 24/8/1958
Photo: Jack Holloway

Lift-off of the first liquid fuel rocket in South Africa. The high-energy igniter leaves an inverted smoke mushroom Photo: Jack Holloway

Cracking the sky

The nosecone is just appearing out of the smoke and fire. The rocket is on its way to 1,6 km, 'the elusive mile' – 28/9/1958 Photo: Jack Holloway

The two-stage rocket one tenth of a second after lift-off – 28/9/1958
Photo: Jack Holloway

Another tenth of a second later and the second-stage motor has ignited, finally sending the rocket to its design alltitude of 1,6 km – 28/9/1958 Photo: Jack Holloway

Final adjustments are made to the igniter. First two-stage rocket in South Africa. My brother David has the worried look! Photo: Jack Holloway

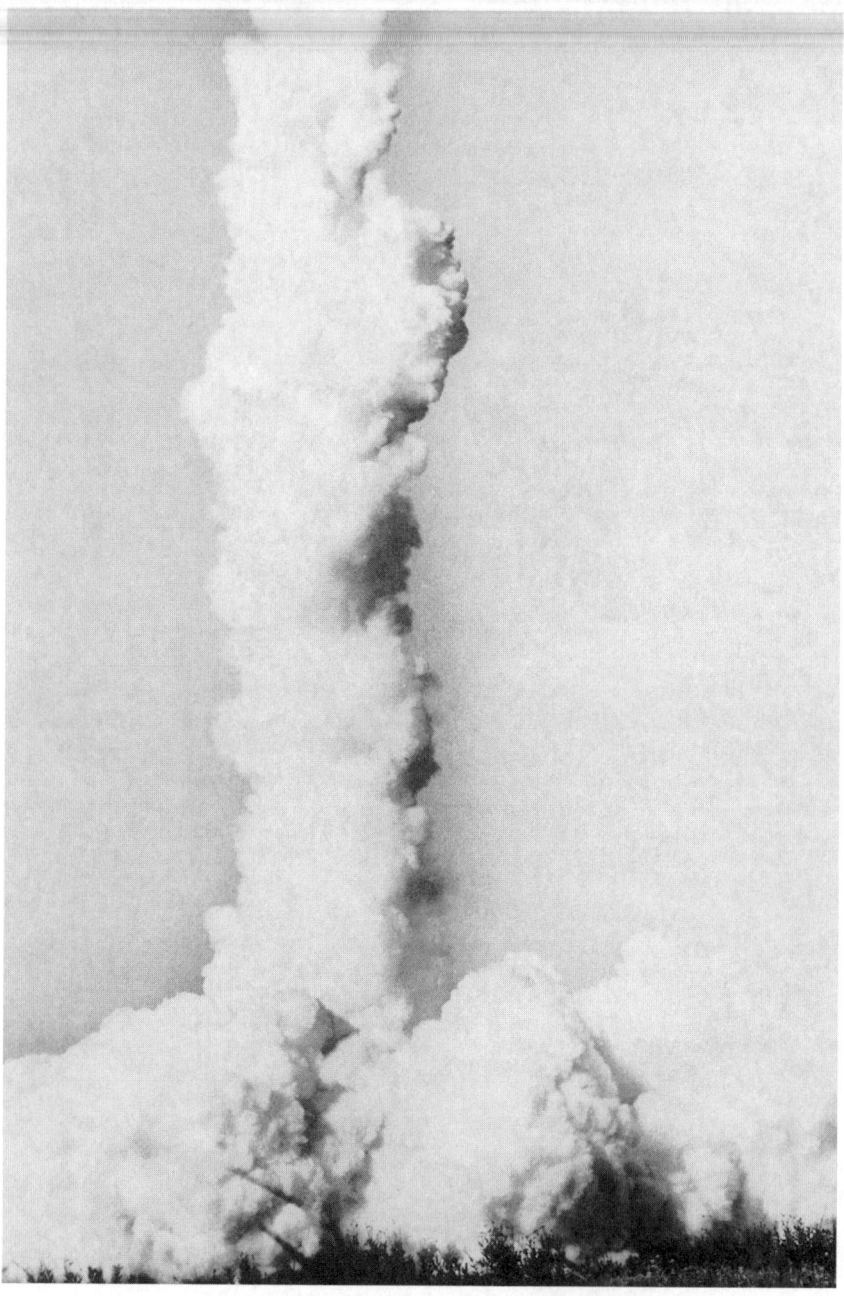
Take-off of the five-stage rocket. Debris can be seen caused by the exhaust blast of the powerful first stage   Photo: Jack Holloway

Sometimes a little help was needed to reach the instrument compartment
Photo: Jack Holloway

Desmond Prout-Jones, dressed in the asbestos suit
Photo: Jack Holloway

The photo gallery

The first recorded rocket to be fired in South Africa – 1953   Photo: Jack Holloway

Static firing of a large solid fuel rocket motor. A down-graded motor of this type was used in the last rocket I launched   Photo: Jack Holloway

Second-stage ignition taking place over 1 km in altitude
Photo: Jack Holloway

David and myself checking the wiring at the pad
Photo: Jack Holloway

Saxon 10 detonating on ignition
Photo: Jack Holloway

Cartoonist Bob Connolly of the *Rand Daily Mail* may have had a prophetic caption for this!

Static test of the 'new' fuel. The motor is secured so that no movement is possible
Photo: Desmond Prout-Jones

Our first attempt at firing a three-stage vehicle
Photo: Jack Holloway

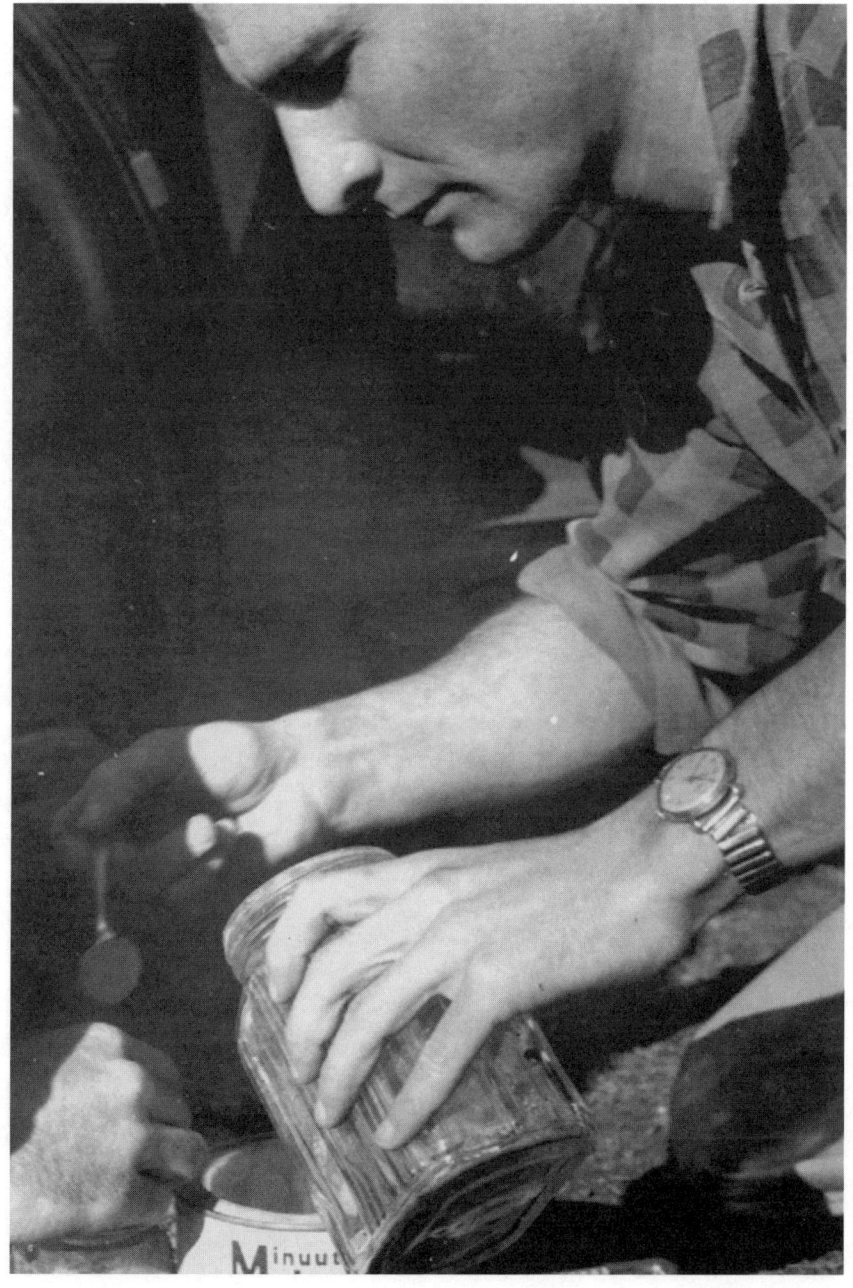

Preparing the ignitor for Saxon 18
Photo: Jack Holloway

The author with a rocket he designed and launched in 2001. It reached 1 km
Photo: Derek Prout-Jones

work on the exhaust nozzles to be finalised.

To obtain the desired shape of these components meant hours of precision lathe turning and Jack thoroughly enjoyed himself completing this task! Verniers and micrometers were the order of the day as the nozzles were being fashioned. They were the heart of the rocket motor's designed performance and if they were not accurately made, the motor would not attain the rated thrust or worse, if too small, cause an explosion. Here Jack's patience came to the fore. He revelled in this type of challenge and usually came out on top. With this new rocket he had taken on five times the normal amount of work. During the days and nights that were required to produce these items he could be found bent over his machine, quietly humming to himself. And might luck be with you if perchance you happened to interrupt him!

The other necessary hardware had been generously donated and assembly of the fins to the various stages was being undertaken at the same time. However, the rocket frame would be completed some time before the nozzles and special bulkheads were ready. The rocket in its unfinished form was sent for degreasing and spray painting at the factory of one of our sponsors. This factory did an excellent baked enamel finish on the previous 12 rockets and before the black lettering or spin markings were applied, the all-white rocket had the appearance of a major electrical appliance!

At last the awaited nozzles and spring-loaded bulkheads were finished and the installation into the various stages was undertaken. When finally assembled, the rocket towered into the air! The trim tabs on the first stage were the first to be used by us and were intended to hold the long airframe steady during lift off. These tabs were part of the fins and could be adjusted to any angle if required.

This rocket was an amalgam of many ideas and designs from the past that had proved reliable. With a vehicle this size no chances could be taken, as we now had five motors to ignite and, more frighteningly, five pieces coming down! And these pieces were not all going to fall in the same place! Their impact points were carefully calculated and safety was uppermost in all our minds! Imagine a finned object landing in your lounge without an invitation! I had to make sure of my figures and I went over them many times to ensure that nothing had been overlooked.

I approached every newspaper that had attended our launchings to assist me in finding a greater piece of land on which to fire this rocket. They willingly obliged by running articles accompanied with pictures. The main thrust was to the farming community and possibly to well-off businessmen who owned large tracts of ground. The response was an overwhelming zero! Not a solitary offer was received!

While the press were running the request, I wrote a letter to the Department of Defence stating my predicament as well as the objectives of the group. In the letter I suggested that perhaps the artillery range outside Potchefstroom in the western Transvaal would be an ideal location. The written answer I received was short and to the point!

'No military property was to be used for civilian enterprise.' So there it was in a nutshell! Not a soul or organisation was prepared to step forward to offer any assistance. So much for the advancement of science!

The five impact points of the various stages were spread along a path 2,4 km from the pad when launched at an angle of 89°. I was happy with this, as our range was quite capable of handling this distance with absolute safety. Now enter the naughty boy! I had no intention of divulging the expected altitude of this rocket to anyone other than members of the group! No press. No photographers. From here on I would do my own thing and to heck with the consequences. Therefore no pictures, so no proof!

I abhorred doing this but I felt justified by the manner in which I had been treated by the authorities and the general public. If the government inspectors happened to arrive while I was getting the rocket ready, I had already worked out that I would fire the rocket. But as a two-stage and then as a three-stage. Quite legal! My request to the inspector was simply that new ignition systems and fuel were going to be put through their paces! Not altogether untruths!

I had clipped the official consent to my clipboard as usual and began running through the pre-flight checks I had drawn up. There were many more pages than normal because of the complexity of the vehicle and I had read through the schedule of events dozens of times. Finally I was satisfied that nothing had been left to chance and made copies for the rocket crew.

The fuel for the upper stages had been mixed and poured into the rocket motors as soon as the hardware from Jack was finished. This fuel required at least four days to harden and set. The fuel for the two lower stages would be mixed at the range on the day of the launch and loaded into the rocket. Certainly the aspect that drew most attention from the group was that only one igniter would set the whole show on the road! Days before the scheduled launch, members of the group had arrived at all odd times to see this rocket standing in my back garden. I had secured it so that it could not fall over and it really impressed the boys. They had seen it in its embryo stage but not in the completed assembly and as one wag put it:

'Why don't you put someone on it? It sure is big enough, then we can claim that we have also given a person a rocket ride!' I treated that remark with mock indignation!

The extension to the launch tower was a lot more work than first imagined. The launch rails had to be lengthened by almost double their original length to ensure a guide at lift-off, as well as adding more supports to stabilise the structure. Remember that the pad itself had been firmly concreted into the ground when we first used this range. Being immovable, all the alterations had to be done at the range, so this meant transporting the necessary welding equipment as well as all the material and odds and ends to the pad.

The days that we could spare away from our full-time jobs were spent in pleasant sunshine out in the open. It is strange that when one is occupied with a task that one enjoys, time goes into 'fast mode' and literally flies by! The time spent working on the tower was passed with good-humoured banter and friendly insults being hurled around by all and sundry! Needless to say, all good things must come to an end and the pad and tower finally reached completion. There was a hint of sadness as the tools and equipment were packed away, but I felt a sense of pride looking at our new launch facility. This was the result of my friends sharing in what I believed in and it made me feel very humble.

One day to go and I started to feel the old butterflies! I tried to gain solace in the thought that the others were perhaps feeling the same as the time grew inexorably closer. Would the countdown reach 'Ignition'? And, more important, would the first stage behave as designed? This was the motor that would impart the tremendous thrust to move the entire rocket up and out of the tower at sufficient speed to add stability to the vehicle. In the design of this stage I had purposefully increased the fin area and also added the trim tabs to slightly overdamp the stability at the lower velocity at launch. I tried in vain to push these thoughts from my mind as I concentrated on my work at hand.

The following day was a repeat of the previous. At this rate I would end up on the funny farm! That particular Friday never seemed to reach 13:50 and the afternoon seemed interminable, but at last it was time to head home. On my arrival I was surprised to see that three men were waiting for me on the porch. They were members of the group and had come directly to my home before setting off for the range! How enthusiastic can you be? It was comforting to know that they had experienced the same excitement as I was.

The loading of the rocket into the vehicles was especially carefully carried out as the three upper stages were fuelled and any bump was to be avoided. This was not for fear of them igniting, but an undue impact might cause cracks to appear in the hardened propellant. This could prove disastrous as the flame front could race into any crevice, thereby increasing the combustion chamber pressure to dangerous levels. These stages were supported in a sling to protect them when travelling over rough surfaces on their journey to the pad. As for the first two stages, they could be likened to two lengths of pipe with fins welded to them: the only care taken was to keep the fins aligned.

All of the other required equipment, such as the control panel, the PA system and tracking-receivers, was safe on the back seat of a motor car. The radio apparatus was checked once more before being loaded into the car. We did not want to try and effect repairs to any faulty electronic circuitry at the pad, so precautions were taken to avoid this. Inserting a torch globe in place of the detonator for obvious reasons checked the destruct receiver! My friends had been through these routines dozens of times and they had drawn up their own list of 'things' to take. A fond kiss, 'See you later,' from my wife and we departed into a darkening evening.

The headlights from my car picked out a hare as it scampered off the sand road leading to the bunker and the low bush alongside the road cast strange shadows. Everything takes on a different appearance in the fast fading light of day. The bunker loomed like the long hump of a whale in the distance. I stopped the car at the entrance to unload the equipment I had brought directly into the safe confines of the bunker. My colleagues drew up next to me and set about rigging the lead lights to brighten up the work area. A hundred metres away the launch tower seemed to stretch its rails up to the dark sky. Away to the west, a long streamer of cloud, purple and orange in colour, lay against the horizon. I was filled with joy and wonder with each visit to this special piece of ground and I thought of the coming event.

We had been through the drill to erect this rocket into the launch and it was a sobering thought that we four were not enough! It would take at least another two men to assist in raising it to the vertical. As we worked on the electrical connections we became aware of more headlights lighting up the sand road as more of the crew arrived. A busy time lay ahead for all. For this shoot I had decided that three people should track the flight and not the customary two. One would take a position at the bunker, while the other two would assume their normal stations on either side of the pad.

We finished the time-consuming tasks at the pad and tested the firing and radio circuits with no delays or faults. The tower adjustment to give a launch angle of 89° was set and locked at the base of the pad. We, as a group, had voted for an early lift-off in case unforeseen snags were encountered with this big rocket – therefore the night-time activity to ready as much as we could and leave only the fuelling of the lower stages for daylight. To reach the top of the nose housing necessitated the use of a stepladder and this thought sent a shiver through me! Had Gordon remembered to bring it? When he arrived, he parked his van at the pad but I could see the silhouette of the ladder on his vehicle and heaved a huge sigh of relief!

Bless him!

With the completion of the items to fire the rocket, it was time for coffee and rusks then sleep. Sunrise is early in the summer and at five o'clock the launch area was a hive of activity. I started mixing the zinc and sulphur propellant at the pad in the asbestos suit and, as usual, sweated like mad! It took nearly an hour to load the two lower stages and I must have lost nearly ten pounds in body weight! At last the fuelling was finished and the settling time was observed. I needed this twenty minutes to get out of the suit and get into a pullover. I sat and chatted with the firing officer. He had re-checked the panel and was satisfied that all systems were in order.

The time was now at hand to assemble the stages together and erect the rocket into the launch. Because of the length, the last stage was pushed up first, the fourth following it and coupling into the telescopic fitting and so on, until the heavy first stage was in place at the bottom. The three upper stages with the new fuel had their long central hollow canal filled with the zinc and sulphur mix to aid the ignition. The igniter was fitted into the nozzle of the first stage and the final alignment of the trim tabs was made. The rocket was ready.

I ascended the ladder, switched the on-board transmitter on, and looked to the bunker to see if the signal was being received. It was. I connected the last of the wires at the nose and, after climbing down, carried the ladder back to the control point. I entered the bunker and glanced at the panel, all the indicator lights were off. One could almost hear the breathing of the men, the silence was so profound! I asked for external power to be switched on. The firing officer repeated the request and thumbed the appropriate switch to the 'On' position. He nodded his satisfaction as a yellow glow came from the panel. Power from the batteries inside the bunker was now flowing to the rocket, conserving the transmitter's internal power scource.

'The range is clear and we have a "proceed" from Range Safety.' Flight's voice broke the stillness. He continued:

'The thirty second count will begin at my mark.'

'Thirty seconds and counting.' His voice was louder.

'Fifteen seconds.' I inserted the safety key into its socket on the panel.

'Ten seconds.'

'External charging circuit is on.' This from the firing officer. I watched the meter needle rise rapidly then gradually fall. The transmitter was sending a steady signal and the signal strength meter was giving the correct reading. All OK.

'… 3, 2, 1. IGNITION!'

Saxon 27 was on its way! The smoke and fire that burst out of the large first-stage tail were watched in awe by all of the firing crew. Never had such an incredible sight been seen by any of us! Pieces of rock and dust were blown in every direction as the rocket cleared the exhaust duct and streaked into the morning sunlight. Right on cue, there came another loud blast as the second stage ignited. I was just in time to see the first stage disengage and the flash of the second lighting up!

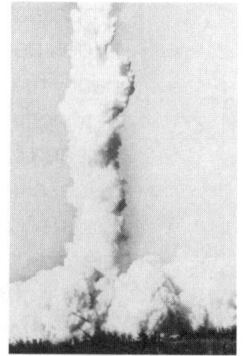

See page 112 for enlarged version

Almost immediately, the third stage sent a spear of flame that kicked the preceding stage spinning away and down. All was happening too quickly to be followed with the human eye and brief impressions were etched into my mind. The third and fourth stages ignited with a short interval between them because of the longer burns. Their sound of life could just be heard away in the blue sky. A sudden double sound like a muted explosion came from above! The sonic boom! We had done it!

I did not see the ignition of the last stage as I sprang back into the bunker to observe the signal strength meter. It was still falling! I stared at it willing it to drop further. The needle began to hover slightly then slowly started to climb up the scale. I marked the lowest point with wax crayon and stood staring at it.

I turned slowly and stepped out into the fresh air, watching the vapour trail descending. It seemed to be a very long way away and there was no sound to herald its arrival. Jack thought that he had captured the lift-off on his 35 mm still camera but could not be certain. He stood next to me and listened to the loud shouts of the men still in the bunker, shutting down all the equipment.

Gordon emerged and, taking my arm steered me to his van.

See page 106 top for enlarged version

'You designed it, now let's see if you can find all the bits and pieces.' He was indeed a practical man. He was right, of course, as all the bits had to be recovered but before we could leave, the tracking crew arrived in a state of excitement.

'Have you any idea where you sent that to?' The three of them had computed the angles and were standing with broad grins on their faces. I returned their smiles:

'Yes, I do. About 24 km, according to my info.'

'It reached 24,8 km as measured by us!' The chief gave me his best smile.

See page 99 for enlarged version

'But did you hear the double bang?' he enquired, with a look of pure innocence.

'That, my man, was the sound barrier being bust!' I answered.

I gave the altitude as 24 km and all were happy as can be. Three of us climbed onto the back of the van and drove off in search of the fallen stages. These would prove the theory of ballistics! As indeed they did. All five were recovered almost in a straight line, with slight deviations due to the uncoupling when the stage ahead ignited. So some were to the left or right of the intended flight path. The final stage was found just over 2,3 km from the pad, buried up to its fins in the ground. It was a total write-off.

See page 105 for enlarged version

We returned to the rest of the men, bringing the spent rocket casings with us. They gathered around the blackened and buckled pieces of metal in silence, remembering what that rocket had looked like standing on the pad. To me, it was a bitter-sweet experience. My report on Monday was to confirm that the objectives had been met and that the new propellant worked better than expected. This report was placed in a file with all the others and I walked out of the inner sanctum a happy chappy!

Chapter 8

Even the gods cheered

I sat alone with my dreams. It was a Sunday afternoon and my family were inside the house. Outside, the sun was warm and as I sat on the front porch, visions of the various rockets I had launched returned to haunt me. It had been three months since the last one and I felt agitated at the prospect that it may have been the last. Such a thought was unacceptable and I grew more ill at ease with the hand I had been dealt. The space race between the USA and the USSR was well and truly under way. The Russians had placed a dog in earth orbit and the Americans had shot Alan Shepard 115 mls into the very boundaries of outer space in answer to the Russians sending Yuri Gagarin on a single orbit around the earth in1961. The game was on and there were no rules. The Americans had a series of failures that put the opposition ahead, as far as the development of a rocket capable of lofting heavy weights into orbit was concerned.

I had read these accounts over and over and with each reading I felt a pang of deep regret that we were not going to be part of this grand exploration of space! Anyway, not for the foreseeable future. A lot of interest had been generated in South Africa with the advent of Commander Shepard being sent on his sub-orbital mission, but at the back of a lot of minds was the belief this was the best the USA could achieve. But it was not enough! In fact, a bit of ridicule was directed their way, even by their own press when things went wrong.

From experience, a rocket had to work 100% or it was a failure. There was no middle road, it was a one-shot expendable vehicle, and hundreds of components had to function perfectly. But from each failure came more understanding and better technology, not only in rocket science, but in other fields of academia as well. Why could our authorities not see this and take a firm stand to advance our own knowledge? This very aspect ate into my soul.

It was with thoughts such as these that I sat outside on that Sunday afternoon, thinking about all those fiery chariots that I had sent upwards towards a greater, yet unknown environment. An endless void that held the key to our very existence extended for light years all around our small, blue planet. The vacuum of space would be conquered, of that I was sure, and a human being would step onto the moon to further our knowledge of the universe. The rocket was the only means by which humankind could break the bonds of gravity and venture to the distant planets, and perfecting this means of travel would take time.

As I saw it, the major stumbling block was not the vehicle but building a reliable pressurised suit to keep a human being alive in that hostile environment. Without doubt, a rocket would be built in the near future to meet this mode of transport, even if it had to attain a velocity of 11,2 km/s! This is termed the 'escape velocity'. The velocity required leaving Earth behind and venturing into the unknown, just as the early explorers on our globe had left the safety of their homes and sallied forth on the high seas. For to discover is to learn.

In September 1961 a radio space research station near Hartbeeshoek was established by the Council for Scientific and Industrial Research (CSIR) in collaboration with the United States National Aeronautic and Space Administration (Rosenthal). This was used for satellite tracking and radio telescope observations with great success. Other installations such as this were founded at strategic spots in South Africa to monitor satellites and deep-space probes. This was our contribution to the space race and could have been of immeasurable benefit to this country if it had undertaken its own rocket development. The scientific means for verifying the performance and position, possibly, of our own space probes were up and running at no extra cost to the taxpayer! What a pity that this advantage was not grabbed with both hands!

But in years to come we would rely on foreign agencies to provide the vehicles and today I am most thankful that I had no crystal ball into the future. For the CSIR had approached me in late 1962 in connection with my rocket research. After showing the representative of that establishment photographs and press cuttings of my activities in rocket research, I was told that the rockets that I was firing were far too big for their purposes! This came as a bitter blow! I was told that I was over-qualified, as the projectile they were looking at was very much smaller. I explained that to downsize a rocket was not difficult and after much debate, the meeting came to an end. 'We will be in touch,' but I heard no more.

These images and sounds ran through my mind as I sat on the porch and my wife's voice brought me to reality as she handed me a cup of tea.

'My, you were miles away in thought. What were you thinking about?' Her enquiry was gentle and not probing. I merely held out my hand to her.

Indeed, what was I thinking? Let the past go and look forward to the happiness around you! I gave myself a mental shake and concentrated on the small birds hopping about at the bird bath

But the fire and smoke would not clear from my mind. When I lay in bed at night I was aware of the wonderful sights I had been so lucky to witness over the years. From the wooden thing that had brought the wrath of my family down on me right up to the big five-stage rocket that was burned forever in me. These were the thoughts that kept me awake at night. Slowly, over time, a germ of an idea started to take shape.

The launch pad out at the range was still there minus the tower. A coat of paint to cover the rusted spots and a phone call to Gordon for the tower and we could be back in business! I had a flash of the chief inspector's face if and when I presented him with my new proposal. It was not a happy sight! I somehow had the feeling that he thought I had by now come to my senses and had forgotten all about rockets. The SARRG was no more and life for him went on without the bother of a rocket lunatic.

As I sat reflecting on things past, one rather memorable incident came to mind and it brought a smile to my lips! Two years previously, a well-known figure in cinder track speedway racing contacted me. Apparently, a special race meeting was to be held to raise funds for children's charities and the organisers were pondering what to present to attract spectators for this worthy cause. The answer: a rocket-propelled racing bike! Enter yours truly!

I met these worthy gentlemen at the track and asked many questions as to the weight of the bike, weight of the rider, average speed that was usually attained and anything that might be relevant. The answers came fast and enthusiastically. A further meeting was arranged for me to fit the two solid fuel rockets to the forks at the rear wheel. Two toggle switches fitted on the handlebars would be used to ignite the motors. The mechanical fittings were ready well ahead of the scheduled date and I ran through the procedure as to the lighting-up of the rockets with the rider, a young man named Doug.

The track was oval in shape and I suggested that as he came out of the first bend, he ignited one rocket. He would then have time to control the bike into the next bend. On coming out of that bend, ignite the second rocket and really head for the finish! All very simple, as he was to be the only rider on the track when this event was to take place. I did not want anybody or thing behind him when those little babies ignited! They shot a flame at least 10 m long and the white smoke could not be seen through. So nobody else was to be present on the track when Doug made his historic ride.

The Saturday night for the event arrived and the spectators crowded into the stadium to the smiles of the organisers. I could feel the impatience in the grandstands as the preliminary races got under way. The interval came and went and at last ... 'Ladies and gentlemen ...' The show was about to start! The motor vehicle with the scraper in tow smoothed the cinder track in readiness for the anticipated exhibition. A tremendous roar went up as, slowly, Doug made his appearance on the bike with two red rocket tubes bolted onto the rear forks. He did a slow, single lap around the track to the tumultuous applause from the crowd. Slowly he approached the starting wires and brought his machine to a stop.

The revving of the motor bike's engine seemed quieter than when a real race with four bikes was about to begin. The wires flew up and the bike leaped into the air as the front wheel came off the ground with the rapid acceleration. He roared into the first turn and I held my breath, waiting for the ignition of the first rocket. As he straightened the bike, there came a huge, vivid flash and a great pall of smoke shot from the rear. Almost immediately the loud roar of the motors filled the stadium. He had ignited both rockets at the same time! I closed my eyes. At a speed far in excess as to what he had expected, Doug could not control the bike as it entered the bend!

He struck the crash barrier with a noise that could not be heard over the din of the high revs of the engine and the last gasps of the rockets! Poor Doug had hit both switches at once and had jumped ahead by some

48 km/h more than he could handle. The crowd was silent as the first-aid men rushed to his assistance but broke into cheers as he was led away on very unsteady legs. I followed and felt dismayed at seeing him stretched out on his back.

Thankfully he had only suffered concussion. When I visited him in hospital the next day, he was able to tell me that he had disregarded my advice and wanted to see what both rockets would do! Oh, rashness of youth! No cuts, nothing broken but, man, what a ride! Needless to say, my nervous system was shot to bits. No more tricks! The rocket motors I had used were safe and reliable and I had no doubts in my mind as to the safety of their function. From then on I kept my focus on the blue above, but the upshot of this caper was that I received more invitations to lecture on rockets in general!

I had a warm feeling inside me as this recollection played itself out in my mind. Yes, there were more things between heaven and earth! The germ of the idea was taking shape and the mists were clearing from my mind as it came now into focus. I was going to design a rocket that would rise into space and present this to the chief inspector and await his reaction! When he saw my proposal he would surely see that no more frolicking on the range was envisaged by me. This was going to be a shot in the arm for rocket research in South Africa.

I spent the following three weeks going over the calculations and working on the drawings of the biggest and most ambitious project I had undertaken. When finalised, the rocket would stand majestic at 13,5 m tall with a take-off weight in excess of 4 t! The maths worked out that an altitude of 180 km could be anticipated! Here was the space rocket of my wildest imagination. It comprised three stages and would burn potassium perchlorate and polyurethane. The thrust at lift-off was calculated at 10 t, more than enough to hurl the rocket into space on a vertical trajectory, reaching into the very region of the gods.

Many details had to be overcome before I dared show the authorities this project. I agonised over the area where such a vehicle could be launched and the same answer came back to me time and again. Over the Indian Ocean! In those days, the obvious place was St Lucia. It was not developed and could be reached by a small convoy of cars that would be necessary to transport the rocket and its ancillary equipment to this place on the KwaZulu-Natal North Coast.

I estimated that the timeframe to complete the construction of everything would be in the region of four months. My group were literally over the moon when I showed them the drawings and calculations! Needless to say, all offered their particular skills to see this bird to the launching pad! I contacted the maritime authorities to hear if I was about to contravene any of their laws or statutes. I was requested to submit, in writing, my objectives at least two weeks before the intended launch date to ensure that no shipping would be passing that area.

So far, so good. My primary concern was the operation of the large first-stage booster. Nothing compared to this beast! Because of the size and power of this rocket motor, I considered it of utmost importance to have a static run and to assess the performance. It was agreed upon to go ahead and build it when I suggested this to the group.

The configuration was to be four 15 cm tubes strapped together with a length of 4,6 m, each tube with its own nozzle and bulkhead. Each motor would produce 2,5 t of thrust and at burnout, the second stage would be ignited, burning the same propellant.

The whole approach to the design of this rocket was a complete breakaway from previous designs in as much as that the fuel was to be ignited at the head end of the rocket and not at the tail. With this type of ignition, the flame was concentrated in a hollow cylinder that ran the length of the fuel charge and burned from the centre out. This had *two* distinct advantages over the 'cigarette burning' fuel configuration:

- The thrust increased with time of burning as more propellant area became exposed as the flame burnt toward the metal wall of the motor. The sides and head-end of the charge were coated with a substance that precluded these areas from ignition. This gave rise to higher thrusts in shorter burning times. In the end-burning configuration, the burning surface remained constant and lower thrusts over longer burn times are achieved.
- The flame front was kept away from the metal wall until the instant of burnout; therefore the motor ran cooler.

The motor was built in accordance with the original drawings and the nozzles were a nightmare of lathe work. One of our sponsors offered to do this, thus relieving Jack of this task. In fact, the nozzles were turned out during two night-shifts on turret lathes and the bulkheads were made in a day. The material used for all components was cold rolled steel bar. The coupling for the second stage was fitted to the centre where the four bulkheads met. This complete unit was to be bolted to a concrete slab at our range. Eight 2,5 cm bolts secured the motor to an angle iron frame on the pad in such a way that if the rocket tore loose it was aimed at the ground where it could do no damage.

We all knew that if the rocket tore loose in the upright position it was anybody's guess as to what damage it could cause, therefore the downward fitting. Even this caused me to double-think the holding bolts' efficiency. They were placed at strategic positions along the big motor, fastening it to the strongest parts of the pad structure. If things went wrong, not only would the rocket be lost but also the pad! A concrete slab was poured into a hole and measured 1,5 m thick and 3 m square. This test set-up was located some 10 m from the launch tower. The metal support structure had four legs each 0,5 m wide and 2,54 cm thick and buried 0,5 m into the concrete bed. If the motor ran as predicted, most of the force would be directed at an angle that would assist the pad in restraining the violent thrust from moving the concrete anchor. Two heavy cross members of angle iron in the form of an inverted 'V' bolted to the pad completed the test stand. The test motor was fixed to this.

Fuelling this motor was very costly, but the chemical firm that sponsored the fuelling never batted an eyelid when I approached them a week before the up-coming static. All that was said was 'Good luck'. Such was the manner in which my 'magnificent five' treated me. Taking the chemicals home was a ride to remember! I feared for the rear springs of my car!

With this amount of chemicals in my possession I drove straight to the range, where I met three members of the rocket crew who had brought the motor and the necessary mounting materials. I certainly was not going to lift that lot by myself! After off-loading the drums, I started to mix the liquid polyurethane and perchlorate crystals. To ensure a constant mix, large quantities at a time were added and thoroughly stirred. Each rocket barrel had a mandrel placed down the centre of the pipe, around which the propellant mix would flow forming the hollow centre section. The mandrels were coated with ordinary shoe polish to prevent the fuel from adhering to them and this should facilitate their easy removal when the mixture had set.

This mixing and loading took many hours to complete, but at last all four motors were fuelled and stood upright in an upside-down position to allow the thick liquid to settle correctly and evenly. The motor was fastened to the 'V' frame in this position and left with a tarpaulin covering the exhaust nozzles.

I knew that it would take at least two days for the cast propellant to harden and set and I had no intention of leaving such a dreadnought at my home for that time. Therefore it was decided to leave it in its fully fuelled state at the range under the watchful eyes of the security people that owned the land. It was a comforting feeling to know that it was safe.

The days passed quickly and the weekend arrived, bright and sunny. A good omen? At 08:00 the rest of the rocket crew arrived and the motor was inspected for any obvious cracks in the fuel at the nozzle end. There were none, so the mandrels were slowly removed and we all held our breath while I did this. All four came out easily and the hollows were clear, all the way to the bulkheads. Everything seemed well and we proceeded to erect the motor into the correct position and insert the bolts into the angle iron frame. This was easier said than done! It took four men to lift it and another to struggle to slip the bolts into the holes located in the frame. The washers and nuts were slid onto the bolts and the whole affair was tightened. The large spanners used would have done justice to a locomotive engineer!

Finally it was secured and I climbed a ladder once more to insert the igniters into the headends. For this, two electrical wires were connected to the 'electrical match' and a charge of zinc/sulphur encapsulated it. The ignition devices were pushed firmly against the bulkheads and a wad of putty was pushed against these to hold them in place when the fire command was given. The wires jutted out of the nozzles and were to be connected to the cable leading to the control panel.

This attempt had no flight implications so therefore I did not submit any request for permission from the authorities. Our safety regulations were in place and I was extremely thankful for the bunker! I had no illusions as to the power that this motor was capable of and every precaution for its safe operation had been implemented.

Each of the firing connections had been made and the red flag was flying on the mast. The gate to the property had been secured with a length of chain and a padlock to keep any unwary visitors out. As there was to be no flight, the tracking men were not required to perform their function but were present to witness this static run.

I held a stopwatch in my hand to time the period of burning compared to the calculated time. The firing panel was the same one that was used for flight firings but for obvious reasons not all circuits would be armed.

'Ready when you are.' Gordon sounded strange, as if the old nerves had at last got the better of him! As he usually timed the flights as well as calling the countdown, he was to time this run as well. I nodded to him. The faces of my companions looked strained in the half-light of the bunker. Were they as stressed as I?

'X minus two minutes and counting.' The count had started. The next few minutes would tell whether we really had the technology to build big rockets! The firing officer shifted his feet and the scuffing sounded loud in the exaggerated silence.

'... 3, 2, 1. Ignition!'

Four white puffs of smoke shot out of the nozzles as the igniters flared into life. A sheet of flame that roared and crackled then settled into a stabbing, waving foxtail of light streamed from the exhausts. This was not the sound of a rocket! It was the sound of a thousand drums, not synchronised, pounding at our ears! I had started the watch at the emergence of the flame, more in fright than by plan and I stared in wonder at the straining motor, shouting its defiance at being held captive. The flame tip was at least three times longer than the rocket tubes and the shock waves were clearly visible in it. Behind the flame a galloping wall of smoke raced across the veld, obscuring it. No other sound could be heard over the tremendous rumbling from the pad and at last it stopped! Smoke drifted upwards from the tail, into the clear morning air. The west section of the range was covered with a massive cloud, moving silently with the light breeze.

The run was over and the motor and pad were in one piece! We sat outside the bunker, each with our own thoughts.

'Gordie, I clocked it at six comma nine seconds. What did you get?'

'Seven on the head. It ran the longest we have had.' His voice was back to normal.

If the burn seemed quick, it must be remembered that with the hollow core, the propellant burning area increases with time and, correspondingly, the thrust increases. I had calculated a burning time of six and a half seconds and we timed it that closed I was a very happy man as I sat and watched the smoke cloud reach the main road, miles away. The wind had broken it up and a tiny segment was still being moved – all that remained of the drama that had played out only minutes before.

The packing-up took almost as long as the setting-up! It was late afternoon when the last of the crew left, leaving me alone at the range. The launch pad with its added pieces stood stark against the sky, a silent sentinel that had witnessed our every attempt. Had we accomplished the things that we had set out to do? Only time would be the final judge of our efforts. Time to go home to the comfort of a family and to tell them how the 'big fella' had performed.

On Monday morning I made an appointment with the inspector and presented myself at the appointed time. I had the feeling that this was going to be a turning point in my life! When I entered the office there was no sign of friendship, only business. We shook hands and immediately settled down to the project in hand. I opened the drawings for his perusal and watched his face closely. These were the drawings I had made for our shot at space – the big three-stage rocket whose motor we had just tested. As his eyes scanned the paper, he pursed his lips and imperceptibly shook his head. I politely enquired if there was anything wrong? He did not answer and continued to look at the drawing. Finally he lifted his head and held out his hand. This was to receive the calculation sheaf of papers and definitely not to shake my hand!

He turned to the last page and stared at the final line of figures. This was the expected altitude. I explained that the maritime authorities had no problem when I told him that I was planning to fire over the sea. I went on to tell him of their requirements and that we would meet them with pleasure. Also, the Department of Civil Aviation wanted to be notified of the expected launch date, also no problem.

'If this thing went off course and headed inland, where would it land?' he asked.

'If that should occur, you know that the last rockets have all been carrying destruct packages on board and that we would immediately destroy the rocket before it could enter any inhabited areas,' I replied.

'That was not my question! Where would it come down?' His voice was quiet and controlled as he continued to stare at me, eyes not blinking.

'If left alone, the rocket would follow the normal ballistic parabola and land somewhere near or in Johannesburg, but the likelihood of that happening is so remote as to be almost impossible.' This was the absolute truth with no attempt to hide anything; I was being as open as possible with all questions asked of me.

'Do you have so much confidence in your systems that they will work 100% all the time and especially in this flying bomb? I am afraid I do not share your optimism and I will never give my consent to such a hare-brained scheme!' These words were spoken in a slightly louder tone than previously as he pushed the pile of papers back at me.

'Mr Humphries, the launch angle dictates the path of the rocket and we have never seen one turn 180° from its pre-set course. Also, you have the written reports about the success that we have had with the destruct package on our flights. We have fired 528 rockets and not once did we have recourse to detonating one. We have proved the receiver in the rocket on more than one occasion and it worked every time instantly. I am prepared to install a backup receiver as well as another ground based transmitter if that would make you change your decision. This could be the beginning of the forerunner for satellite launch vehicles for South Africa. Please give us the opportunity to take this step. If you are not happy about the launch area, you nominate the place and we will go there.' I felt as if I was pleading the cause and disliked every second of it. His reply was what I had now come to expect.

'I have given you my answer, so accept it, and if it comes to my attention that you have disobeyed me, there is a heavy fine or imprisonment for this transgression!' He had spoken the final words and he picked up a sheet of paper and began to read it. The meeting and our association were ended forever. The red tape had won!

I went back to work with a heavy feeling, realising those years of work and a lot of money had just been kicked down the drain. Had this whole effort been in vain? I felt like weeping. With one sentence the Chief Inspector of Explosives had burst the bubble, not only for me, but for my colleagues as well. So be it, the tunnel-vision effect! If I had one iota of fear that this rocket would expose anything to the remotest possibility of injury or damage, morally I would never have wasted my time in trying to get permission. But I knew that because of the launch site, the various stages of the rocket would fall harmlessly into the sea and that each stage would be destroyed if an erratic flight appeared to be starting. Although the third stage would be well out of human sight, the radio signal in the nose housing would indicate whether the rocket was keeping to its vertical flight.

The signal strength meter would fluctuate, indicating that a different path was beginning. At the very first sign of this, the destruct button would be pushed and the rocket destroyed. These pieces too would end up in the sea. The prime target of rocket research is safety, not only for the men that fire them but also for the public and their property first and foremost! I honestly believed that I followed that code of conduct to the letter from my earliest effort. Out of all the static and flight firings that I had supervised, not one accident, to man or property, had materialised. It was not something to be proud of but an absolute necessity! Every branch of research has its own inherent dangers and therefore all precautions are taken to preclude such dangers. With rocket science you are dealing with what is in reality a controlled explosion and if negligence or carelessness creeps in, the control is lost. The result is a violent termination of the test.

I had seen these explosions and they had made me all the more cautious in my approach to every aspect of the design, building, fuelling and launching of rockets. We had a fine safety record but all the time we sought safer ways to accomplish safer procedures. For this, no fault could be laid at the door of the SARRG.

Chapter 9

Lifted on tongues of fire

Life went on as usual, but the sting of the words spoken at the last meeting did not abate. The large rocket motor that we had built resided in my garage and it pained me to see it lying on the workbench. Two other stages had been partially built and they stood in the corner opposite the main stage. For weeks I saw these grounded birds on my every entrance to and exit from the garage and slowly a form began to take shape. I would be in serious grief if I launched that behemoth, but nothing had been said about a smaller version! Herein lay the shape of things to come!

For the next few weeks I sat and sketched the layout of a smaller three-stage vehicle that was not as powerful but would be my swan-song to rocket design. This one was going to be just for me and what better way than to send it higher than the last? Most of the hardware was lying in the garage and with a few modifications; a new rocket would emerge! And only a few souls would know about it and the authorities were not on the list nor were the media. It was to be by invitation only and I explained my reasons to the group when we held the next meeting.

Agreement was reached, but where was it going to be launched? Ah, the same old problem.

'Let us try west this time. I believe that the weather is fine and there are miles of nothing but sand all around.' As I said this I felt it was the right decision. If anything, our range would be covertly watched and any activity seen to be taking place just might be correctly construed! A vote was taken and west it was!

Within a couple of days, work on modifying the rocket had begun. First, the four-chambered first stage had to be converted into a *two-chamber* stage with the length reduced to 3 m and the coupling for the second stage altered. This was to accept a smaller adapter at the nozzle end of that stage. Next, the fins of the first stage were welded to the body, as were the fins of the two upper stages. The last stage had the instrument housing fixed aft of the nosecone. The completed stack stood just slightly more than six metres and weighed a modest 142 kg. The design altitude was 50 km with the first two stages burning the new fuel and the final stage using our zinc/sulphur mix. This mix was easier to ignite and hence more reliable.

The launch date could not be set for a host of reasons, the main one being the construction of another tower that could be transported to anywhere. The final details were not insurmountable, and as the days passed the rocket began to take on the shape that I had drawn as well as the tower taking shape in Gordon's hands. This pad would have to be hammered into the ground to support the weight and not move the guiderails at lift-off. These criteria were overcome and the base of the pad was similar to the one that was concreted into the ground at our range. Stout angle iron braces formed the framework on which the rest of the pad was bolted. The guiderails were slid into lengths of steel tubing of the correct diameter and were one metre long to give support to the rails. As mentioned in an earlier chapter, the rails were disconnected from the pad at the end of each test to prevent theft. They were brought back on the morning of the next test and re-fitted to the pad.

As can be imagined, this assembly was not light and the pieces when bolted together took three men to lift, and that was without the rails! The final fitting of the various components to the rocket was completed at the same time that the tower was assembled for the first time and the rocket stood perfectly on the pad, its twin exhausts pointing directly into the exhaust duct. With the completion of the total hardware, it was taken apart and stored, ready for transport. The tower at Gordon's home and the rocket at mine were both covered from prying eyes.

Two months went by and we stood on the hot sand, shielding our eyes from the bright sun. The sky was a peculiar yellowish colour and the dunes threw stark shadows that contrasted vividly with the low bush that dotted the landscape. We had looked for and found a hard piece of ground where the pad could be mounted. It was assembled and placed in position, the anchoring spikes were hammered into the soil and, when driven in to their maximum, the pad was levelled and tested for rigidity. The guiderails were attached and lashed to the pad to prevent them moving in the strong winds that blew with a sorrowful sound at night.

We had arrived three days before and had set up our tents, a camp in the middle of nowhere. The drive to our destination had taken just over two days before we decided on a bare stretch of ground that showed no signs of life in any direction. It was an ideal piece of land to launch our last rocket. Of the normal 15 members that attended all the firings, only 7 could make the long trek because of work commitments. We had left with the blessings from those who were unable to join us and we sorely missed them. This meant that certain jobs had to be split and shared by those present.

I had the discomfort of mixing the liquid and solid propellants in a tent, out of the wind. It took one day to load all three stages and set them aside to cure. The zinc-based fuel would really have time to settle whilst the first two stages hardened. It was during this period that the pad was sited and all electrical checks were being finalised. As always, nothing was being left to chance and I was going to act as the flight officer in the absence of Gordon and also as the firing officer. Some of the chaps had brought along their own cameras on the understanding that these pictures were for their own use and not for publication. I personally wanted no memento of this launch or rocket, to remind me of what might have been.

For very obvious reasons the exact area where we were about to fire this rocket had to remain undisclosed just as the actual launch was kept from public knowledge. It was tested in secrecy and I did not want it splashed across the newspapers as it would draw unwanted attention and publicity. This is all that the authorities were waiting for – to see if I would disobey them. Well, I had not, but at the same time I could do without another tongue-lashing. This rocket was certainly nowhere near the size of the one I had originally planned, but if it worked we would know that our design and systems were functionally correct. This would mean a great deal to know that our work had not been all for naught!

The time for launch was set for early the next morning and we turned in early to be awake at dawn. I had spent the afternoon calibrating the radio equipment and marking the signal strength meter. As well as carrying the radio aloft, a delayed smoke pyro was placed in the nose. That would ignite when the third stage started its trip earthwards and would continue to leave a smoke trail for tracking. The trackers would follow the flight with high-power binoculars and should see the smoke trail easily.

There was a moment of anxiety when the rocket was loaded onto the pad as the aerial from the transmitter snagged and almost pulled loose. This took 15 m to put right and the rest went without incident.

We had no bunker, so Mother Earth was our next best. The control panel was placed on the ground and the crew lay flat on their stomachs. From where we lay the 100 m of cable to the pad looked like a winding crack in the earth. The trackers had left to take up their positions on either side of the pad and we awaited their signal that they were ready. We had brought along a three metre rod and attached the yellow flag to one end and the red flag to the other. This would notify them when ignition was about to take place as the rod would be held aloft, flying the red flag

Final checks completed, I commenced the count. Heat hazes danced crazily in the distance, giving the effect that I was looking through a sheet of water. The slender shape of the white rocket on the pad appeared to move slightly as the heat rose from the dry sand. I shook my head to satisfy myself that it was an illusion.

'T minus one minute and counting.' My palm was wet with sweat as I held the stopwatch at the ready. One of the rocket crew had volunteered to operate the panel and I gladly handed this task to him. His name was Arthur, he was one of the old hands, and I had every trust in him.

'Master?' I read from my own pre-launch procedure on my clipboard.

'On!' came the rapid reply.

'Power?'

'On!'

'Transmit?'

'On!'

'Heater?'

'On!'

'Rocket clear!'

'… 3, 2, 1. Ignition!'

I had one ear listening to the 'beep' of the transmitter and I lost its sound as the first stage blasted into life. I thumbed the stopwatch into ticking. The rocket was lost in an enormous dust cloud, only a dim flame showing as it sped away and upwards. The loud crackle and roar beat down on the desolate, low scrub sending a noise that had never been heard before, shattering the silence. Dust was rolling in towards us where we lay and I cupped my hands around the signal strength meter, trying to shield it as best I could.

The first-stage roar ceased abruptly only to be followed by the crackle of the second-stage ignition. This noise rapidly diminished as the rocket screamed higher. A distant bang as the last stage fired, immediately accompanied by a double boom! I kept my eyes on the meter and was surprised to see how quickly it got to 30 km! I had an overwhelming urge to twist the set-screw on the face of the meter to the 50 km mark! I shook myself mentally and concentrated on the needle, dropping slower now but never did it once waver. The flight and lift-off had been textbook stuff!

I held the crayon ready to mark the glass front of the meter as soon as the needle stopped and here it was – 48 km! It did not reach the 50 km altitude as planned and I would never know why. A slightly shorter burn on one of the stages? A slight delay in ignition of the second or third stage? The descending smoke trail plummeted to the ground and suddenly the point of it and the ground were one. The recovery team drove off, sending up another dust cloud in their hurry to get to the impact point.

I had stopped the watch when I had marked the meter and glanced at it now – 3,3 min. I did some quick arithmetic and found that the rocket had an average speed of just over 872 km/h its maximum speed was an estimated 1 700 km/h.

The recovery boys brought the three stages back and the final stage had landed a little over 5 km away, well and truly dug into the ground. I walked to the pad and sat down.

The rest of the men started to take the tower apart and load it onto the van then the remains of the rocket. To me it was a depressing moment, knowing that this was the last time that I would have the privilege of seeing one of these vehicles rise into the blue sky. I relived the nervous tension as the count was reaching the moment when the event was irreversible and the grand times I experienced with my fellow rocket men. There was no rush now and soon the aroma of coffee filled the air. All unnecessary equipment and utensils were stowed in the vehicles, so we sat and enjoyed the silence that sang in our ears. I had the distinct impression that I heard the gods laugh!

The journey home was uneventful and seemed to last forever. I pulled into my driveway and turned the car off. It was early evening and I could not wait to share the happenings with my family. It was good to be home with the loved ones that had been my support and encouragement for all that time. Bless them!

I phoned the men that had accompanied me to make sure that they had all arrived safely and was relieved to hear that they had. It had been a long, hard trip to launch this last rocket, but the satisfaction that I got from seeing my dreams come to life was well worth all that had preceded it. I had learnt more in almost 15 years than most men would, and that knowledge was something that would live with me for however many years I would inhabit this planet. *Lift-off*!

Chapter 10

A look at history: fire that moves

It has long been established that the Chinese invented gunpowder in the thirteenth century and inadvertently discovered the rocket at the same time. One of the ingredients, saltpetre, was found in abundant natural supply in China and India, and was used as an agent to cure meat. It is likely that this white powder was found to have other uses when, perhaps, a little spillt onto the fire and ignited! As wood was the main fuel for fires, the saltpetre reacted with the hot, burnt wood that we know today as charcoal. This was the second necessary compound. Whether sulphur was later added to this composition is uncertain, but what is known is that sulphur resin was later used in 'Greek fire' that was catapulted in earthen pots. This additive effectively prevented the defenders from easily extinguishing the flames, and at the same time gave off copious quantities of a noxious fumes. The arrows dubbed 'Chinese fire' were used mainly as incendiary devices against besieged cities.

Possibly, the first indication of the propulsive effect came as the result of this flammable compound being loaded into bamboo rods, lit and, not having the rear tightly sealed, whizzed off with fire streaming from it.

Thus it is possible that as time went on the 'fire-arrows', as the incendiary hollow bamboos became known, were used in the Battle of Pieping against the Mongols in AD 1232. In AD 1242 Roger Bacon became the first European to successfully compound a mixture of gunpowder, and his formula was not written in the plain Latin of the day because of the importance of the mixture. It was regarded as a secret.

Bacon has not been accepted as the inventor of gunpowder as its use was well known in the East long before he experimented with it. But by keeping a strict watch on its composition he played a large role in its potency.

The first mention of rockets in Europe appears to be in AD 1258. They were used in the battle for the Isle of Chiozza in 1379. Further early development was due to a German engineer, Kyeser von Eichstredt. He experimented with rockets in 1405 using gunpowder in different proportions. Fifteen years later, in 1420, an Italian, Joanes de Fontana, is recorded as having carried out similar tests.

By the mid-fifteenth century the rocket had become a fully-fledged weapon. It was used in 1449 and again in 1452.

Towards the end of the century, rockets were used against cavalry and also for signalling at night. A written account by Hanselet in 1630 refers to a type of grenade carried by a rocket for use as a light shrapnel shell! Experiments were carried out in Berlin in 1668, when shots with rockets weighing up to 120 lbs proved successful. They were made out of strengthened wood cases to withstand the pressure and heat. The largest rockets carried a warhead of 16 lbs.

In the book *Asiatic Researches* (volume 3), published in the late eighteenth century, an account of a battle at Paniput in India in 1761 states that the Rohillas fired such great numbers of rockets in salvoes that the enemy, the Mahrattas, could not bring their horses under control and could therefore not charge – no doubt because the noise of the rockets was totally unexpected and also because the chaos they caused with the troops.

One can only imagine the disarray!

During the Indian campaign, rockets were used extensively against the British cavalry towards the close of the eighteenth century. These rockets had an iron casing 8 in long by 1,5 in in diameter, with a spiked nose, and used an 8 ft rod to balance them. Special 'rocketeer' troops launched them.

When British troops fought against Raja Hyder Ali of Mysore in 1780, they ran into a new threat. Hyder Ali had improved upon the pasteboard casing and constructed larger rockets up to 12 lbs in weight and fitted with a 10 ft pole. These projectiles were capable of ranges in excess of half a mile but were no more stable than their Chinese counterparts. To make up for this, the rockets were fired in salvoes of dozens to hundreds. This drove the British off the battlefield at Guntur in 1780.

An artillery expert, Sir William Congreve (he was knighted in 1814), was fascinated by the prospects of the rocket as a bombardment weapon and he started work on a new casing with an improved mixture. As a result of his experiments, Congreve was given access to the Royal Laboratory at Woolwich where he made several military projectiles. These proved to be so successful that they were used during the European Wars of the early nineteenth century

Incendiary rockets were designed and played an important role in the fall of the French harbour base at Boulogne on 8 October 1806 during the Napoleonic Wars.

They were also responsible for the destruction of a portion of the French fleet. The rockets were fired in salvoes from 24 specially constructed boats that were conveyed to the scene of battle by larger vessels. The rockets used were specially developed for marine warfare and contained a liquid incendiary that was ejected on impact. This liquid emerged from a number of holes drilled around the pointed nose and on exiting immediately covered the target with fire. These weapons were used against wooden ships of that period and were able to inflict untold destruction. The pointed nose ensured that the rocket penetrated its target, releasing the deadly fire!

Congreve had gone out to see the effect his rockets had under conditions of hostile action against the French. They were used again in action in 1807 at the siege of Copenhagen. Perhaps this is the best-known action where the Congreve war-rocket was used. Over 25 000 rockets were used in a real fire barrage and the city was completely destroyed.

The projectiles weighed 32 lbs and were capable of carrying incendiary, explosive or a combination of both, varying from 8 to 20 lbs, to a maximum range of 2 mls. Thousands of these rockets were expended in this assault. But they were far from being accurate at this period, as they were unstable in flight and therefore volley application yielded the best results. When Congreve replaced the guiding rod with fins, the rocket started to show some advancement in accuracy. However, this improvement cannot be credited to Congreve as a French artillery engineer, Frezier, had written about the adding of fins to rockets in his technical book on armaments some years earlier. In the same work, Frezier also foresaw a later refinement of Congreve's, the 'two-charge rocket', to which I shall refer later.

A notable contribution to the science of rocketry prior to the twentieth century was the invention of the axially rotated rocket. This eventually saw the end of the rod- or fin-stabilised projectile. In this configuration, developed in America in 1815, the rotation was brought about by having small multiple nozzles spaced around the main nozzle at an angle and causing an effect similar to a Catherine wheel (Gatland & Kunesch 1953). This improvement caused much interest and rocket enthusiasts around the world aimed their efforts at improving this new innovation.

In England, Congreve produced many rockets similar to the American version. He also imparted spin by applying offset thrust using angled nozzles.

Congreve left designs for rockets up to 8 in in diameter and weighing 500 to 1 000 lbs. He passed away in 1828, but two years earlier he had patented a system with a series of rocket cases laid in line within a single tube. After the first had been ignited, the rest fired in sequence, thereby increasing the burning time and range. This method is what Frezier had foreseen. This layout was the forerunner of the multi-stage rocket and unwittingly had laid the foundation for the vehicles that would one day leave the embrace of Mother Earth!

Although the rocket had been used mainly for military purposes, another role was awaiting it. This application was for a humanitarian operation! John Dennet of Newport designed rockets weighing 23 lbs that had an iron case and an 8 ft rod for guidance. In 1826 these rockets were put to use as a life-saving device when a number of life-line stations were built at various points on the coast of the Isle of Wight. These rockets had a range of some 250 yd and Henry Trengrouse had experimented with modified Congreve rockets in 1807 to use them for this very purpose.

He had successfully fired a line across a small bay at Porthleven, and so demonstrated the viability of the rocket in this new role. However, his demonstrations were turned down because of the advent of the Manby mortar system that had officially been adopted in 1806 by the coastguard authorities. In 1842 Courte developed a similar system, also using modified Congreve rockets, and shortly afterwards, Dennet brought out a design using two rocket cases strapped together. The range was increased to about 400 yd, but difficulty was experienced in igniting the rockets simultaneously and differences in thrust yielded inaccuracies of flight.

It was only in 1855, after Colonel Boxer had developed a two-stage rocket of long range, that the rocket finally came into its own and replaced the Manby mortar. Rockets used for this purpose were small in comparison with their predecessors and weighed only 0,5 to 2 lbs. The advantages were readily seen as the rocket launching stands were more compact and hence more portable than the mortar apparatus. More importantly, the line was illuminated by the exhaust and could be followed at night.

They were adopted everywhere in 1870 and since that time have helped to save many thousands of lives around the coast of Great Britain alone. The 6 lbs rockets of that time were capable of carrying a light hemp line about 1 000 ft in length. The rocket passed over the ship, allowing the crew to secure the line. A heavier line was attached to the line on shore and this was then pulled to the stricken vessel. Personnel and passengers used the heavy line to make their way ashore using a breeches buoy slung from the heavy line.

William Hale, an Englishman (1797–1870), who devised the first spin-stabilised rocket, designed a rocket that was to be utilised by the American and British armies. This rocket appeared in the British Army's *Treatise on ammunition* in 1905. Two versions of this projectile were used in the US Army: 2,25 in diameter weighing 6 lbs; and 3,25 in diameter weighing 16 lbs; the maximum ranges were 1 760 yd and 2 200 yd respectively. However, the days of rockets as bombardment weapons were numbered, as the search for more sophisticated weapons continued. The emergence of the breech-loading cannon with a rifled barrel gave an increase in range plus improved accuracy, and largely replaced the war rocket.

A book entitled *Half hours in air and sky* (Anon 1899) surprisingly gave an accurate account of rocket theory by an unknown author! It refered to the advancement of physics to the point that a machine of the rocket type could be developed to transport people! He went on to add that the rocket did not rely on its exhaust stream pushing against the atmosphere and therefore could operate in a vacuum. The human being was the drawback because of the necessity to breathe air, but with such a machine we could leave the boundaries of our planet and journey to other 'orbs'. What foresight! Even today the principle of rocket propulsion is misunderstood and yet this author of long ago had envisaged the journeys that we, a hundred years later, are striving to undertake.

The first person to suggest seriously that the rocket could be used as the vehicle for space travel was a Russian, Konstantin Tsiolkovsky, the son of a forester. He is credited with developing the basics of rocket propulsion and proposing the use of liquid fuels. He published a paper in 1903 expounding his theories. He concluded that for space travel, the spaceship must have a sealed cabin with oxygen and a system for air purification. He theorised about the medical effects on humans of weightlessness in space and all the most pertinent aspects as to fuel quantities and the speed to overcome the force of gravity. Tsiolkovsky is regarded as one of the great pioneers of space travel, but did not receive recognition until late in life. The launch of Sputnik on 4 October 1957 took place a month after the hundredth anniversary of his birth.

In America, Dr Robert H Goddard (born 1882) was working on similar theories without the knowledge of what Tsiolkovsky and others were doing. In the same year, 1907, a French engineer, Robert Esnault-Pelterie, was preparing a mathematical paper into the practicalities of space flight. It was submitted to the Societé Française de Physique and published under the title *L'Astronautique* in 1928. It is regarded as one of the classic studies. In 1907 Goddard also prepared a paper suggesting that heat from radioactive materials could be used to expel substances at high velocities through a rocket motor to provide sufficient power to enable flight in the vacuum of space.

Goddard was ahead of his time and actively went about trying to practically prove his theories. He was the first of the early pioneers to combine scientific insight with practical experiments in propulsion. In 1919 the Smithsonian Institution published Goddard's paper 'A method of reaching extreme altitudes', which contained information and conclusions from the tests he had carried out. He stated that a rocket could be designed to reach the moon using the step principle, and explode a load of flash powder to herald its arrival on the lunar surface. The American press latched onto this and Goddard was thereafter known as the 'moon rocket man'. Goddard was a shy man and this derision made him much more reticent about his work. He went on to prove that, contrary to popular opinion, rocket thrust was effective in a vacuum. He very soon became aware that liquid fuels were the most suitable propellants for taking a rocket beyond the Earth's gravitational field.

On 16 March 1926 he conducted the world's first liquid-fuelled rocket flight in Auburn, Massachusetts. The rocket was propelled by liquid oxygen and gasoline and reached an altitude of 41 ft. The historic début lasted 2,2 s! In total, it covered a distance of 184 ft at an average speed of 64 mph. The rocket was a lightly constructed affair with the motor at the top and the two propellant tanks beneath. Pipes connected the fuel to the motor above. The total length of the apparatus was 10 ft. The idea was that with the propulsive centre being ahead of the centre of gravity, the thrust would naturally pull the rocket straight up!

Goddard soon realised the fallacy of this line of thought and he wrote:

> this is evident from the fact that the propelling force lay along the axis of the rocket, and not in the direction in which it was intended the rocket should travel, the condition therefore being the same as that in which the chamber is in the rear of the rocket. The case is altogether different from pulling an object upward by a force which is constantly vertical, when stability depends merely on having the force applied above the centre of gravity.

From 1930 to 1942 he worked at Roswell, New Mexico, and was financed by the Guggenheim Fund for the Promotion of Aeronautics. Here, his rockets achieved altitudes of over a mile and he experimented with gyro-controlled rockets. This mild-mannered professor was the father of the liquid fuel rocket and from 1942 to his death in 1945 he worked for the US Navy at Annapolis, Maryland. It is incredible to think that from his small beginnings it took just a couple of decades to develop the giant Saturn V rocket and land a man on the moon! Robert Hutchins Goddard must be smiling, wherever he is, at having seen his dream come true!

This was the virtual end of the individual in rocket research and organised groups of amateur rocket enthusiasts took up more intense activity. A small group of German enthusiasts met in Breslau in June 1927 to found a society devoted to the study of space-travel. From this meeting emerged the Verein für Raumschiffahrt (VFR), which means Society for Space Travel, and had as members such notable men as Herman Oberth, Willy Ley, Rudoph Nebel, Max Valier, Johannes Winkler and Wernher von Braun: all to become contributors to rocket development. Winkler became the first man in Europe to fire a liquid-propellant rocket. Professor Oberth, of Hermannstadt, Transylvania, published a small, but highly technical book, with the title *The rocket into interplanetary space,* in 1923. His theories fired the imagination of engineers in Europe as to the possibilities of space flight, with the result that further books appeared, mainly supporting Oberth's statement, although it had brought him ridicule and criticism from more orthodox engineers.

The French pioneer aviator Robert Esnault-Pelterie abandoned his favourable aviation career to devote himself to astronautics after World War 1. Among his many contributions he demonstrated the possibility of inertial navigation guidance for spacecraft by means of self-contained, automatic devices. Between 1930 and 1940 he searched for the ideal fuel, and conducted some experiments, but with the invasion of France in 1940 his work was temporarily stopped. He continued his research into the promotion of spaceflight after the end of war.

The work of the foregoing scientists set the platform. Significant achievements depended on the development of unprecedented multi-stage rockets that could take instruments, animals and, eventually, humans into space. The call was for remarkable engineers capable of designing and building such rockets; and two men answered the call: a Russian, Sergei Korolev and a German, Wernher von Braun, first in Germany then the USA. These were two exceptionally brilliant men whose engineering and organisational capabilities turned concepts and theories into real projects and rockets.

Korolev was born in 1907 in what was then Russia and is now Ukraine. In 1930 he left flying school to concentrate on rocket development and in 1931 he directed the Group for the Study of Jet Propulsion. This organisation was both a design and manufacturing centre for rocket prototypes. Under his direction the Soviets built their first liquid-propellant rockets. But under Stalin many engineers and scientists were arrested by the Secret Police and imprisoned. In 1938 Korolev and aircraft design engineer Andrei Tupolev were imprisoned and forced to work in a scientific labour camp. With the German invasion of Russia in World War II, the camp was moved from the Moscow area to a place east of the Ural Mountains, where they were out of reach of the Germans.

At the end of hostilities he returned to rocket engine development and he was assigned to coordinate the assimilation of German rocketry expertise. As chief designer of guided missile development, he was able to organise a section that produced the first successful Soviet intercontinental ballistic missile in August 1957 and two months later, on 4 October, launched Sputnik. This was the world's first artificial satellite and the international propaganda success allowed Korolev to proceed with his visions about launching a human being into space.

The then Soviet leader, Khrushchev, recognised that the ability to launch satellites, probes, and later people into space supported his political and military challenge to the USA. In less than a decade after Sputnik Korolev had achieved many firsts in the 'space race'.

First man in space, Yuri Gargarin; first woman in space, Valentina Tereshkova; first 'space-walk' by a man, Aleksei Leonov. The Soviet Union successfully crash-landed a probe on the moon and took the first photographs of the far side of the moon with Luna 2 and Luna 3 in 1959.

Called the father of the Soviet space programme, Sergei Korolev's contribution to the development of space exploration was enormous. His identity was kept secret and he was only known as 'The chief designer of launch vehicles and spacecraft'. Only after his death in 1966 following surgery was his name made known to the world.

The other giant of rocket research was Dr Wernher von Braun. He was born in Germany in 1912, and early in life revealed his leaning towards science and music. In the mid-1920s he became an enthusiastic member of the VFR and rocketry that would occupy him for the next 50 years of his life and bring him worldwide fame. It was he and his teams that designed and fired the successful V-2 rocket during the war, against which there was no defence. Only massive bombing raids carried out by the Allies on the factory at Peenemunde brought the dreaded threat to an end.

At the end of the war Dr von Braun and top members of the V-2 project handed themselves over to the US Army. They were taken to the Redstone Arsenal in Huntsville, Alabama, to begin work on a long-range ballistic missile, but before this transpired, von Braun and his scientists used captured V-2s to teach Americans about rocketry. This took place at Fort Bliss in Texas, as well as tests at White Sands, New Mexico. Their work progressed and in 1956 they fired a Jupiter C missile that travelled 3 300 miles and reached an altitude of 680 miles. This was a three-stage rocket and again von Braun turned his attention to space exploration.

He got the go-ahead from the Secretary of Defence and prepared a satellite launch vehicle. This rocket was called a Juno 1 and was a modified Jupiter C with an added fourth stage carrying the satellite. The first attempt at placing a satellite in orbit had been given to the US Navy, using their sophisticated Vanguard rocket. The whole world watched this launch via television and were horrified to see it lift slowly from the pad and, equally slowly, settle back again amidst a huge fireball! Now it was von Braun's turn to fire his Juno 1!

On 31 January 1958, the rocket roared off its pad at Cape Canaveral, Florida, and soared into the dark sky. For anxious minutes, scientists listened for the radio signals that would indicate success. The rocket carried its payload into a higher altitude than was planned and as a result it took eight minutes longer for the satellite to travel around the world. At last the telemetry signals from Explorer 1 were received, announcing that, indeed, it had gone into orbit! Wernher von Braun had become a national hero!

The National Aeronautics and Space Administration (Nasa) was established in October 1958. In July 1960 von Braun's group became the central core of the George C Marshall Space Flight Centre, in Huntsville. This was Nasa's major centre for the development of rocket launch vehicles that included the Saturn 1,1 B and V. The engineering and managerial complexities involved in developing these huge rockets are a testament to his capabilities. Outstandingly efficient, each of these rockets was launched on schedule and performance and safety were achieved. The Saturn V rocket became an integral part in the endeavour to land a human on the moon.

The development of the Saturn V rocket was the focal point in his career with Nasa. He remained director of the Marshall Space Flight Centre until 1970, when he was transferred to the planning office in Washington. In 1972 he retired from Nasa to the private industry to focus on other efforts to promote spaceflight. He was awarded the National Medal of Science along with many other honours during his distinguished career. He passed away in 1977. Wernher von Braun played a prominent role in the development of rocketry and spaceflight.

Chapter 11

Fire that moves: the rocket

The advantage of solid propellant rockets is their inherent simplicity because of the lack of valves, pumps, controls and other moving parts. They are easily utilised and do not require any servicing. A typical solid fuel motor has the following components:

- Propellant charge, sometimes referred to as a 'grain'
- Combustion chamber
- Igniter
- Mounting fixtures to locate the unit if not an integral part of the vehicle
- Exhaust nozzle.

Modern-day propellants have a rubbery consistency and burn on their exposed surfaces to generate hot exhaust gases that flow through the nozzle and produce a reaction force or thrust.

Solid rocket propellants must contain both an oxidiser and a fuel. Composite propellants consist of a heterogeneous mixture containing a separate oxidiser such as one of the perchlorates in a resin binder with a metallic fuel, such as aluminium powder. Homogeneous propellants such as nitrocellulose contain the fuel (carbon and hydrogen) and the oxidiser (nitrate groups) in the same molecule. Propellants consisting primarily of a mixture of nitrocellulose and nitroglycerine are known as 'double base' propellants.

Once the propellant grain is ignited, it burns uniformly on all the exposed surfaces in a direction normal to the burning surface.

Definition: The thrust of a rocket is equal to the product of the mass flow rate and the effective exhaust velocity. If a large thrust is required, then a large amount of propellant must be consumed in a short time. This can be achieved by having a large burning surface area or a fast burning rate, or both. Conversely, a propellant with a small burning surface can deliver a lower thrust for a longer time.

In addition to varying the shape of the grain, it is also possible to prevent some surfaces from burning by coating these surfaces with an inert material or 'inhibitor'. In a simple end-burning rocket the burning surface does not vary during operation. In most other shapes there is some variation of burning area with time.

There are three types of thrust variation with time. If the grain is designed so that the burning area and therefore the gas evolution rate, the chamber pressure, and the thrust increase with time, then the rocket is said to have progressive burning characteristics. If the grain is designed so that these parameters decrease these quantities, then the rocket is said to have regressive characteristics. A grain that maintains approximately constant burning surface and thrust is said to have neutral burning characteristics. In my experience the progressive grain has proved to be more efficient in the applications to which I subjected it. In practice, as the time of operation increases, the thrust increases, the chamber pressure increases, and the heat of combustion is kept away from the chamber wall until the instant of burnout.

A 'cigarette' burning grain has the outer cylindrical surface inhibited and burns down its length. The exposed burning surface therefore remains constant (or it may increase if a core is formed).

If the thrust decreases, then the most likely cause is nozzle erosion, which will lead to a drop in pressure and a decrease in the burn rate. (Propellant burn rate increases with increasing pressure and this magnifies any change.)

The combustion chamber is usually cylindrical in shape and is sealed at the head-end by a closure, known as the *bulkhead*. This prevents the hot gases from escaping through the front of the motor. Grains that burn radially outwards from a hollow core are not necessarily progressive. At present it is common practice to use a 'star'-shaped port designed to give a roughly constant surface area evolution and therefore a constant thrust. The igniter is fixed at the bulkhead and, when activated, shoots flame down the hollow core of the grain, igniting the charge from top to bottom.

To understand this type of grain more fully it is necessary to go to the mixing stage, right up to ignition. The organic fuel is in a liquid state and the oxidiser powder is stirred into it until a uniform mix is obtained, adding a little oil as a plasticiser. A steel mandrel, the length of the chamber, is inserted into the empty chamber up to the fixed bulkhead. A template is attached to the exhaust nozzle to keep the mandrel perfectly straight and in the centre. The mandrel, with a light coating of grease, is positioned and the liquid poured into the chamber, filling it completely. The filled chamber is set aside vertically for at least two days to allow the mixture to harden. Once cured and hard, the mandrel is extracted, leaving a hollow right through the propellant grain. Lightly greasing the mandrel means that it is easily withdrawn from the chamber, leaving the desired grain configuration. An inhibitor is located at the bulkhead and nozzle ends to prevent the flame from starting between these two components.

The igniter, as mentioned, is fitted to the bulkhead at the front of the motor to ensure that the entire hollow charge is ignited at the start of the test run or flight. The igniter must give off enough energy to ignite the propellant instantly.

In the 'cigarette' burning charge, the igniter is placed at the tail end of the motor and the flame burns the exposed circular section upwards toward the *bulkhead*. The old life-saving rockets and modern skyrockets basically employ this system.

Certainly the most critical component in the solid propellant rocket motor is the exhaust nozzle. The design of this single piece of hardware will determine the operation and efficiency of the motor. Nozzles are of the convergent and divergent type, the hot reaction gases being compressed through the convergent section. As this high velocity, hot jet stream enters the divergent section it rapidly expands, permitting a decrease in pressure and an increase in velocity above the velocity of sound. Additional expansion in the nozzle exit adds to the thrust of the burning gases. If the nozzle terminated at the throat section, the gas exit velocity would be sonic.

These nozzles are named after their inventor, De Laval. If the chamber pressure is too low for a specific propellant, ignition may occur and immediately be extinguished, only to start again and stop. This condition is known as 'chuffing' after the sound that is produced. To ensure smooth starts, I fixed a 'burst diaphragm' enclosing the nozzle exit. This diaphragm enabled the pressure and the temperature to rapidly increase before it ruptured, allowing the propellant charge to ignite fully. This condition was only encountered with the larger, higher thrust motors.

The specific impulse of solid propellants is perhaps the single most important property regarding performance parameters of solid rockets. It may be defined as the thrust that can be obtained from an equivalent rocket that has a propellant weight flow of unity'. The equation is:

$$Is = F/w$$
$$= c/g$$

where Is is the specific impulse in pound of thrust per pound per second of propellant flow, F is the thrust in pounds, w is the weight flow rate in pounds per second, c is the effective exhaust velocity and g is the gravitational constant.

Before any calculations may be undertaken in motor design, the specific impulse of the proposed propellant must be known. Without this information, realistic design cannot be attempted. Other parameters that are essential in the design of solid propellant motors are:

Burning rate	= r (symbol)
Specific weight	= pb
Specific heat ratio	= k
Chamber pressure	= pc
Desired average thrust	= F
Maximum vehicle diameter	= D
Desired burning duration	= tp

The above examples are intended to show in a simplified form a few of the design criteria before the actual design is tackled. For each propellant category and application, a different design approach is needed. Furthermore a good part of the basic design information is based on previous experience and available experimental data. Today this data is available in numerous excellent books, and the would-be designer is well advised to procure and read what is entailed before setting out into the experimental field.

Classic examples of successful large solid fuel rockets are the boosters on the space shuttle the Minuteman and the strap-on boosters of the Delta that launches deep-space probes. Possibly the best-known solid rockets are the two boosters on the sides of the Space Shuttle. Each is 149 ft long and holds 1,1 million lbs of solid propellant. When ignited, along with the three liquid fuel main engines, a total of 3,3 million lbs of thrust is generated to lift the shuttle from the pad. They burn for a little more than two minutes and are discarded at burnout to parachute into the sea, where they are recovered for re-use. The abbreviation for these solid rocket boosters is SRBs.

It is worth noting how to arrive at the throat diameter. Whether it is for a solid or liquid fuel combustion chamber, the equation is accurate. First, we have to determine the desired pressure that the chamber must run at, and from this we can calculate the nozzle area:

$$At = F/cfpc$$

Where At = throat area
cf = thrust correction factor
cp = chamber pressure.

This calculation assumes that the convergence/divergence angles of the nozzle are 15° and 30° respectively. This configuration has proved to be the optimum in most designs and the throat area will be in square inches.

For the thrust calculation, the following is used

$$F = c(w/g)$$

Where F = thrust
c = effective exhaust velocity
w = weight of propellant flow
g = gravitational constant.

The thrust calculation, as above, will calculate the thrust in pounds, as the metric system was not in use when I applied these formulae. Metric units can be used, however. (I still get confused when not using the old feet and inches!)

I always used a minimum safety factor of 40% when designing the combustion chamber as I considered safety far more important than a bit more thickness in the wall of the chamber. As I mentioned in the early chapters, the solid drawn steel tube that I used had a burst pressure of 1 500 lbs/in^2. But I never designed the motor to run at more than 1 000 lbs/in^2. Thus a safety margin was built in from the start, for to see one of these motors explode is something one would only like to see once!

A little more on the system that I employed to ignite the upper stages of multi-stage rockets. Initially, a pullout electrical plug on the side of the rocket was anchored to the pad by means of a pre-measured length of piano wire. When the plug was pulled, contact was made and electric current fired an igniter in the next stage. The drawbacks can easily be seen, as the wire was prone to being burnt away before the rocket had reached sufficient height or the plug did not disengage. Sometimes it worked, but most times it resulted in a misfire and a lot of money up the Swanee!

To overcome this unreliable method, a spring-loaded valve was devised that would open when the chamber pressure dropped below a certain value. The coil spring was located outside the bulkhead and exerted its pressure against a square piece of hardened steel plate, via a hole, inside the chamber. At ignition of the propellant, the pressure inside the chamber greatly increased, slamming the plate flat against the hole through the bulkhead. This created a positive seal at the head-end of the chamber until near-burnout, when the plate was forced away, opening the port to allow the residual hot gas to escape. This system worked very well with motors of short burning times, as the hot flame did not have time to influence the metal flap. It worked every time and much of the success of our later rockets can be directly attributed to this innovation. Later, with the advent of the microchip, time delays became an easy piece of equipment to make and very reliable. The bulkhead was fixed to the chamber by eight machine screws and was positioned deep enough to allow the upper stage nozzle to telescope into the chamber.

Construction of the Saxon Alpha series of rockets

In the design of these large rockets, only solid drawn steel tubing was used for the motor casing and solid steel bar stock was used for the nozzles. It never ceased to amaze me how much of the solid steel was turned into swathe as the nozzle took shape! Generally, almost three-quarters, and at times more, of the original mass was removed to produce the desired component! This material proved efficient in the relatively short burning times that were expected of the motor. Aluminium, although lighter, and therefore more desirable as far as weight saving was concerned, was mentioned in an earlier chapter when I explained how the exhaust gas eroded the material. It was actually the high temperature of combustion that melted the aluminium. Aluminium was acceptable only in very short burn applications. Even at burning times of less than a second, I saw aluminium nozzles severely damaged.

The fins were cut out of ordinary cold rolled steel sheet and oxy-acetylene welded to the motor tube. The alignment of each fin and its placement were carried out with the greatest precision, as the flight path was dictated by the angle of attack of the fins. In large zinc and sulphur rockets the take-off is exceptionally fast and in the first second after giving the launch command, the rocket rises 800 ft! Therefore if the fin alignment is out, the flight is going to be a doozy! Whizzing where you don't want it to whiz! So we were extra careful in this part of the assembly.

The nosecones were turned out of a hard wood for two reasons: it was easier than forming them out of metal; and they were a natural electrical insulator. The instrument compartments, directly behind the nose, were of aluminium tubing and bolted directly onto the forward bulkhead of the motor. In this section, the radio receiver for the destruct package, the radio transmitter for height assessment and the circuitry for the smoke flare were housed. The antennae for the radio equipment were led out and fixed to the wooden nose. The instrument compartments were screwed to the nosecones. Before spray-painting, the entire rocket and ancillary metal parts were de-greased and passivated to eradicate any unwanted pollution.

It must be stressed that any new design of rocket motor was always static test-fired to evaluate the design and safety. These tests were carried out at the range with the motor under test securely bolted to the pad, thereby preventing any movement. In all, 102 static tests were carried out with no detonations but a few underperforming units were recorded. The motors were always angled towards the ground as an extra safety precaution in the event of their tearing loose from the pad.

The igniters were self-made and consisted of a torch globe with the glass bulb removed. This left the thin filament wire exposed and a thin tape 'cup' was wound around the top of the threaded section, protecting the fragile filament and also serving as a receptor for a small quantity of the zinc/sulphur mix. When the 90 V were applied to this arrangement, ignition of the mix was instantaneous and released enough energy to ignite most propellant combinations. It was a simple and reliable way of achieving reaction to start rocket motors. With later igniters, the commercially available 'electric match' could have been used, but I always used the torch globe arrangement as not once, in 102 statics and 528 flights, had it failed to do the job. If it works, do not fool around with it!

The control panel was described earlier but it is worth revising it here. The panel had six toggle switches, a sprung-loaded pushbutton for launch and, of course, the safety key. This key was in my possession at all times and during that time, the panel could not send any commands to the rocket. Seconds before the intended firing I would insert it into the panel, allowing current to flow into it, and as each switch was closed in the correct sequence, an indicator light came on. The 'heater' switch preheated the igniter, so that when the launch command was given ignition was immediate. Another safety feature that I built into the panel was that if the switches were closed in the incorrect sequence, the current was immediately switched off. No games, please! From an early age I had come to respect this activity and was very strict in the safety regulations that I had my crew adhere to. I believe that it was because of this that no mishaps took place.

Last, and perhaps the most important aspect of each shoot, was the determination of the altitude achieved. Initially, two members would position themselves 2 000 ft apart, in a straight line, from the rocket. They were armed with an inverted protractor mounted on a wooden rod. To this, a piece of cord was attached to the zero degree mark and hung straight down by virtue of a weight at the end of the cord. When the rocket reached its zenith, a smoke charge was ignited and the two men took a 'fix' on the smoke, by placing their finger on the cord and reading off the angle. The tangent of each angle measured was taken, and multiplied by the distance from the pad. This then gave the vertical height reached by the rocket.

As the systems became more sophisticated, a radio transmitter was installed in the rocket. A portable receiver had a microvolt meter connected to the automatic gain control circuit and this meter was practically calibrated in points of a mile. As the rocket flew away from the pad, the signal strength became weaker and this registered on the meter, but in this case the needle indicated distance. The trigonometric and radio results were compared before an altitude was officially given. There were hardly any discrepancies between the two, but the lower result was taken, just to be on the safe side!

An interesting table may be drawn up between the liquid and solid propellant systems, as it is difficult to draw generalised conclusions on their relative merits.

Liquid propellants
- Can be less sensitive to temperature
- Can be started and stopped
- Usually lighter unit for longer burning times
- Often gives higher specific impulse
- Thrust may be varied readily
- Often involves simpler propellant preparation.

Solid propellants
- Simpler in construction and design
- Usually the lighter unit for low total impulse applications
- Few servicing problems
- Believed to be more reliable
- Sometimes difficult to handle.

Each particular type must be very well scrutinised for the optimum utilisation of the inherent characteristics of the two systems. This is the reason that there are different types of launch vehicle today. The first intercontinental ballistic missiles were extremely complex liquid fuel rockets and required set times to bring them to launch status. These times were, for certain types, too long and a simpler, more ready missile was sought. Enter the Minuteman, a solid fuel missile, ready to go!

The solid fuel rocket has worn many guises over the centuries and will continue to do so. To date it has been the answer to heavy-lift vehicles, assisting in launching great weights into high or low earth orbit. Or to give that extra push for interplanetary missions where the liquid propellant sustainer motor needed that little more power that was economically more viable with the addition of the solid fuel boost.

Needless to say, the arena of guided missiles depends heavily on reliable solid fuel rocket motors, where second chances are virtually nil in combat. The Gulf War showed the world just how far these missiles had progressed. In the foreseeable future this type of rocket motor still has a large role to play both in war and the exploration of space.

The liquid propellant rocket

At the beginning of the twentieth century, it was realised that the way to outer space could only be achieved by the use of liquid propellants. They offered higher specific impulses and longer burning times, two important factors in obtaining the necessary velocity required to shake off the shackles of our home planet's gravitational attraction. But these advantages came at a price. The liquids, for the most part, were unstable, and the mechanical aspects were complex. To obtain maximum performance, the design of the loaded structure was anything but simple. For one thing, the mass ratio of the vehicle had to be in the order of 80% fuel and 20% structure. In other words, the rocket was essentially a flying fuel tank! The early 1950s American Viking rocket was in this order that led to the emergence of the satellite launch vehicle, the Vanguard.

The following table is appended to illustrate certain fuel combinations:

Oxidiser	Fuel
Liquid oxygen	Ethyl alcohol
Liquid oxygen	Kerosene
Red fuming nitric acid	Aniline
Hydrogen peroxide	Hydrazine hydrate
Liquid oxygen	Liquid hydrogen

When assessing the liquid fuel combinations, the specific thrust is used instead of the specific impulse, as used for solid fuel rockets.

These are a few of the combinations that have been experimented with and nearly all are noxious or highly corrosive and require special handling with the correct equipment. With red fuming nitric acid and aniline, spontaneous combustion occurs when the two mix. The others require an energetic igniter for combustion to commence.

The basic components for a liquid propellant motor were given in a previous chapter and they are worth considering individually at this stage.

The combustion chamber

The chamber is a special device into which the propellants are introduced, mixed, atomised and burnt at a high temperature to produce gaseous reaction products, which in turn are accelerated and ejected at high velocities. Because of the high rate of energy evolution the cooling, stability of combustion and injection problems must be considered. Note that I am using the term *combustion chamber* instead of the technically more acceptable *thrust chamber* and for this I apologise. From my earliest memories of liquid fuel rockets, I have always referred to it as the former. (Old habits are hard to quit!)

The *combustion chamber volume* is important and this is chosen with attention to the following considerations:

- The volume has to be sufficiently large enough to permit complete mixing, evaporation and combustion of the propellants. If the volume is too small, the propellant combination will have incomplete burning, and this is generally seen in an excessively long exhaust flame. This is caused by unburned fuel igniting outside the chamber.
- For longer burning periods the wall surface area has to be cooled by one of the methods mentioned in the next section. To reduce the cooling requirements it is desirable to decrease the exposed wall area and the local heat transfer intensity. As the chamber volume is decreased, the exposed wall surface area also decreases, but with an increase of heat transfer intensity. This is because a smaller chamber volume necessitates high gas velocities. For any chamber size, geometry, and fuel combination there will be a volume for which the average heat intensity and wall area are a minimum.
- Rockets are airborne vehicles and therefore weight is at a premium, and a light weight dictates a smaller chamber volume. The weight will be a function of the operating pressure and the configuration.
- Manufacturing and design indicate a preference for a cylindrical chamber. Cooling jackets for more complicated forms are difficult to make and design for efficient cooling.
- The maximum diameter of the chamber often determines the rocket's dimensions. A smaller chamber offers lower aerodynamic drag because of its smaller structure.

These are a few of the factors to be taken into consideration where chamber volume is concerned. Others such as gas pressure drop, vibration, coolant pressure loss and types of injector are all consequences of chamber volume. The *characteristic chamber length* is defined as the length a rocket of the same volume would have if it were a straight tube and had the same volume and had no converging section. This is given by:

L* = Vc/At

Where L* is the characteristic chamber length in feet, At is the nozzle throat area in square feet, and Vc is the chamber volume in cubic feet. Here the chamber is considered to include all the volume up to the throat area.

The *stay time* Ts of the propellant gases is the average time spent by each gas molecule in the chamber volume, hence:

Ts = Vc/wV1

Where Ts is the stay time, Vc is the chamber volume, w is the propellant flow rate in pounds per second and V1 is the average specific volume of propellant gases in the chamber in cubic feet per pound. From the foregoing, it can be seen that chamber design is a little more than producing cylinders with fire streaming from the rear!

The *injectors* deliver the propellants into the chamber in the correct ratios and pressures and are different for each application. There are many types of injector but all have the same relevant task to perform. Apart from the two functions mentioned above, the injectors have to introduce the propellant mixture in such a manner that atomisation and mixing will result in a proportioned homogeneous fuel/oxidiser mix, one that can be readily vaporised and burnt. Injectors include spray injection, pre-mixing type, splash plate, self-impinging and shower head. They are to be found at the forward section/headend of the chamber and are coupled to the propellant lines.

There are several ways of delivering the liquids to the chamber, but only two will be dealt with here. The simplest form is to pressurise the fuel tanks with an inert gas, such as nitrogen, at high pressure. The high-pressure gas is stored in a container, usually at the fore of the fuel tanks. Plumbing from the gas container leads the high-pressure gas through reduction valves to the fuel tanks, and from the tanks more pipes conduct the fuels through pressure regulators to the injector bank. At the launch command, the reduction valves are opened and the tanks are immediately pressurised, forcing the liquids through the regulators and into the chamber. At this stage, the fuels enter the chamber under pressure, and because the injectors, mix, atomise, vaporise and ignite. A pyrotechnic device that is fired a fraction of a second before the fuels enter achieves ignition. It is common practice to lead the oxidiser ahead of the fuel to prevent a 'hard start'; this ensures combustion as soon as the fuel arrives. I have experienced this happening only once with one of my own motors! The above description applies to my own designs and the sequence of events that followed the start procedure. As mentioned, the fuels were liquid oxygen and petrol. This motor design is suited to 'small' rocket applications, as we shall see.

Turbo-pump pressurisation

As opposed to the high-pressure gas expulsion system, the turbo-pump serves large rocket motors that require vast amounts of propellant per second. Goddard experimented with them, and von Braun perfected them and used them in his V-2 rocket most successfully. The moon-rocket, Saturn V, consumes 25 t of propellant to start its five first-stage motors! Once the motors have ignited and are operating within their parameters, the consumption drops to 15 t/s! Each of the five motors guzzles 3 t of fuel per second and 2 000 000 kg of propellant are exhausted in 2 min 40 s. The thrust generated by these enormous motors is over 3 000 000 kg. Engineers would be hard pushed to achieve these requirements with conventional gas pressure systems.

The alternative is the turbo-pump! In these assemblies, generally speaking, high-speed single- or two-stage impulse turbines coupled to centrifugal pumps are used. The main design problem is deciding on a concession between efficiency and cost. The turbine wheel follows normal steam turbine practice. The turbine disc and shaft may be forged as a single piece in high tensile steel, or separately, depending on the design. The impulse blades are shrouded and steam enters the blades through a single nozzle. Usually the steam is passed back through the blades by means of guide vanes that serve to provide a form of velocity compounding. The turbine shaft is supported in two bearings: a roller bearing, or a ball-thrust bearing fitted between the turbine and the fuel pump. One is fitted at the extreme end of the shaft outside the pump.

This description is typical of the basic turbo-pump used on small to medium-size motors. The power to drive the pump is obtained by breaking down hydrogen peroxide with a liquid or solid catalyst, such as potassium permanganate. When these two chemicals are brought together in a steam generator, the resultant reaction is super-heated steam at high pressure. This high-pressure steam is directed at the steam turbine and extremely high revolutions of the centrifugal pumps are reached. Cavitation of the pumps is prevented by pressurising the fuel tanks to approximately one atmosphere to ensure that the liquid propellants are at the pump inlets before start-up. The fuel pump in the V-2 had a flow of 56,1 kg/s, a discharge pressure of 310 psia and a shaft speed of 3 800 rpm. The oxidiser pump had a flow of 69,3 kg/s, a discharge pressure of 250 psia and a shaft speed of 3 800 rpm. It produced 465 hp.

Fuel being burnt at this rate produced 25 400 kg of thrust for a little longer than a minute. This rocket was ahead of its time in as much as that the systems, that is the pump, motor controls and steering vanes, were still being used ten years after the end of World War II. The Redstone missile was a modified, elongated V-2 and this was modified to launch the first American satellite!

By comparison, the pumps in the space shuttle deliver the following:

Pressure = 422 psia stepped up to 4 300 psia
then again to 7 420 psia
Speed   = 28 120 rpm.

These figures relate to the oxidiser pump, using liquid oxygen. The fuel pump is somewhat different:

Pressure = 276 psi stepped up to 6 515 psia
Speed   = 35 360 rpm

This fuel is liquid hydrogen. The fuel pump peaks at 76 000 hp!

These figures stagger the mind! Little wonder that the three main engines consume well over a ton of fuel each for every second of operation. Comparable flows may be obtained by gas pressurisation but the high-pressure gas vessel would be so heavy that it would not be practical for any air-borne vehicle. The motive force to drive these pumps is derived from the vaporisation of the cryogenic fuels, and instead of the exhaust going into the atmosphere, it is recycled back into the fuel system.

I could not afford such luxuries, as turbo-pumps in my efforts with liquid propellant rocket motors and, besides, the units were not big enough to economically justify them. When the propellant flow rate becomes of such proportions that a compressed gas system will not practically suffice, then the pump system is viable. Perhaps one day I shall have the opportunity to design a motor, using one of these efficient ways of propellant expulsion. I am only too aware of the complexities involved with regard to the sealing between the two pump housings and the drive shaft. A leak at one of these locations would allow a situation to develop that may easily end in a detonation! Many bangs have happened in the past and I am sure that there are still more to come, as more powerful motors are designed and the pump is the key to supplying the fuels to very thirsty combustion chambers.

The cooling of the combustion chamber is an extremely important design feature of any liquid propellant motor. We have seen the extreme temperatures reached and they are severe enough to melt any metal. In the early days a method referred to as the 'heat-sponge' was employed with some success for motors with short burning times. This method revolved around the construction of the chamber being made with thick blocks of metal to act as a sponge and 'soak' up the heat. When the wall temperature approaches the melting point of the wall material, then the danger of local melting, high internal stresses, and the reduced physical strength properties make a longer rocket run dangerous.

The most efficient way to cool the motor during operation and to increase the burning time is to circulate one of the propellants around the motor prior to injection. This is achieved by surrounding the chamber with a light metal cooling jacket with a small gap between it and the chamber. One of the propellants is pumped into the jacket at the nozzle-end of the motor and completely circulates around it before being injected into the chamber at the headend. This effectively reduces the temperature of the chamber walls as the flowing cold liquid carries the heat away. This method has the added advantage of partially vaporising the fuel, making it more readily ignitable. The vast majority of modern rocket motors utilise this method of combustion chamber cooling.

The sequence of events in starting a turbo-pump rocket is as follows:

- All propellants are loaded. (Oxygen, fuel, peroxide and permanganate.) The tanks are pressurised and sealed to maintain propellants at the pump inlets.

- At the start command, all propellant valves are opened and the pyrotechnics inside the chamber are ignited.
- Propellants enter the pump under the influence of gravity and are fed into the chamber. These are ignited and at the same time the hydrogen peroxide is sprayed into the gas generator, where it decomposes with the permanganate, creating super-heated steam. This steam is led directly to the turbine. The turbine immediately speeds up and the rotating pumps now pressurise the propellants, and the liquids under great pressure enter the chamber through the injector bank, in a fine, mixed spray.
- Ignition of this mixture takes place and the motor spontaneously rises to the designed thrust level. The exhaust from the pumps is directed out of the tail of the rocket via a suitable pipe.

The amount of peroxide loaded should always be a little more than the anticipated burning time. This is to ensure that the main propellants are completely used by allowing the pumps to operate slightly longer. This applies to an expendable vehicle only. In vehicles, such as the space shuttle and the multi-stage Saturn V, the motors must be capable of shutting down and restarting, therefore the fuel quantities for the pumps and the motors are calculated accordingly.

The same approach is adopted for a straight high-pressure gas-fed system where the expelling gas is sufficient to empty the propellant tanks. Again, careful consideration must be given as to the exact quantities of fuels and gas. Too much of either and the result is dead-weight being carried that in turn diminishes the ultimate altitude. Too little of either and the motor shuts down early and the result is the same. This discussion of these two main propellant expulsion systems, I hope, will give a clear insight into some of the problems that are encountered.

Simplified vertical aerodynamics

In a vertically ascending rocket, it is assumed that the Earth is stationary, the thrust axis lies in the direction of flight, and side forces are zero. The rocket is designed with the centre of gravity ahead of the centre of aerodynamic pressure and suitable rear-mounted fins give the rocket arrow-like stability. *Pitch*, roll and yaw are inherent in any body moving through a medium, and in this case it is the atmosphere. Pitch is when the nose of the rocket moves up and down. *Roll is* when the rocket rotates about its longitudinal axis and *yaw* is when the nose swings from side to side. But another factor creeps into the equation – drag. With proper design the first three can be dealt with. Several other quantities, such as control forces and stability aspects, enter the analysis, and since the lift coefficient, drag coefficient, altitude, flight angle, and burning time depend on a number of independent variables, which are different for each vehicle, no general solution can be given. Step-by-step answers are usually indicated.

The drag forces are another matter! If not properly seen in the design and construction of the rocket, these can have serious consequences. They are the most important aerodynamic factors affecting the performance of a vehicle. Movie pictures have been taken of a rocket deviating from its original flight path very suddenly and show the vehicle disintegrating and exploding. It had been subjected to forces far above its design limits and had been completely destroyed. The drag forces that apply to a flying rocket are:

- Friction drag
- Pressure drag
- Interference drag
- Parasitic drag
- Induced drag.

Air molecules bumping into the rough surface of the rocket skin and rebounding and bumping into one another cause *friction drag*. Therefore the smoother the skin, the less friction drag.

The impact of tiny air molecules on an object protruding on the surface of a moving rocket produce *pressure drag* and the area behind the object will have negative pressure. An object the size of a rocket moving at speed through the air simply does not give the air time to slip out of the way and close in behind it. In other words, the nose shoves the air out of the way and creates *pressure drag*.

When the boundary layer airflow is interrupted over the body and the fins by the junction between these two, that is, it is caused by the interference of the airflow between the two surfaces, *interference drag* is created.

Parasitic drag is caused by anything that juts out from the rocket body to cause interruption of the smooth flow of the boundary layer over the vehicle.

Induced drag is created by lifting force, or by the lifting characteristics of a surface such as the rocket's fins. The nose and body combination does not add enough lift to be of any consequence. The greatest induced drag is created at the fin tips. If a fin is at an angle of attack and thereby causing lift to stabilise the rocket and to restore it to zero angle of attack, a high pressure will exist on one surface and low pressure on the other. The high pressure will spill over the fin tip in attempt to relieve the low pressure on the other side. The resultant is a 'span-wise' flow over the whole fin. This is seen predominantly at the fin tip. This produces a corkscrew motion (helical) of the air and is known as a vortex that is shed from the fin tip. Watch a commercial jetliner take off on a rainy day and you can see the vortex streaming from the wing tip. Energy is required to maintain this vortex and this loss of energy shows up as induced drag. Therefore the smaller the vortex, the less drag.

The rocket, in its entirety, has to be very carefully considered at the drawing-board and many calculations have to be made and discarded before one comes close to an optimum configuration. The swept-back fins that appear so good in pictures and comics, giving the impression of high speed, are in fact the worst when it comes to induced drag. Far better the clipped delta shape that has less drag and will not be prone to 'flutter'. (Flexure at high angles of attack can cause the fins to vibrate and eventually break.)

In possibly 90% of the rockets that I flew, the fins were of the clipped delta shape and the flight paths achieved were as planned. No pitch or roll was ever observed, only a slight yaw, which was caused by over-damping. I mentioned that the fins were generally slightly larger than necessary to ensure a stable trajectory. The clipped delta fin form that I refer to is a right-angle triangle that has been cut off at about 5/8 along its horizontal axis.

Different nose shapes were flown and the shape that operated best is referred to as 'ogive'. This nose is not the pointed cone shape that is generally associated with rockets, but a rounded front, almost in the shape of a half a sphere. The air rolls over it cleanly and does not have to contend with any sharp change of direction. All noses were highly sandpapered to give the smoothest possible finish and to reduce drag to a minimum. The body and fins were degreased before spray-painting and once again a smooth finish was the order of the day.

To observe the rocket's behaviour at lift-off, black stripes were painted on the body and fins so as to detect any deviation from the launch position. This was closely scrutinised on movie film of the launch, played back one frame at a time, and it was through this method that no anomalies were detected. The rockets left the tower perfectly stabile, heading straight up. What was seen was that no matter how fast the rocket left the pad, the hot exhaust blast always engulfed the vehicle as it lifted off! The extreme heat from the blowback caused wires to melt and solder joints to simply disintegrate! This led to the addition of the exhaust tunnel under the base of the rocket with no susceptible components outside the rocket. The tunnel was angled to the one side of the pad and the hot exhaust gases blasted into the tunnel and were led away from the base of the vehicle. Generally, most of the smoke and fire could be seen exiting from the left-hand side of the pad when viewed from the bunker. This deflection of the exhaust did not come cheaply as the tunnel had to be replaced after every launch because of the intense heat and erosion. The fitting to the pad was initially seen as a maintenance fixture and was designed to have replacement ducts fitted. The price entered the picture when procuring the metal sheet and forming it into the necessary shape. The fitting of a new duct to the pad was by means of six heavy bolts, which at times were badly burnt and proved difficult to remove.

Those movie pictures were worth their weight in gold, as the cause of certain malfunctions would have not been detected had it not been for them! It became part of the launching to start the movie camera several seconds before the fire command was given. This allowed Jack time to ready the still camera to record the lift-off. There is only one still photograph that captured the rocket's nose just emerging from the fireball. Only after studying the movie film was the seriousness of the situation realised. Fortunately the solution was quick and effective, but the importance of being able to view the cause demonstrated to me what happened in a few hundredths of a second.

One cannot appreciate the awesome power that a well-designed motor delivers and the noise and blast that goes with its normal operating conditions. The 100 m smoke trail of Saxon 18 leaving the pad gives some idea; the smoke roaring out the exhaust still touching the ground; the mighty blast and scattered debris as the large five-stage rocket left the tower and headed skywards, a smoke pillar like never before! Bear in mind that these photographs were taken at one thousandth of a second shutter speed and few actually show the rocket riding atop its fiery thrust.

Lastly, *safety*, this word was our breakfast, lunch and dinner! A rocket of any size is a potential hazard and must be respected as such. The smallest commercial skyrocket can be a minor explosion waiting to happen. If the powder propellant is cracked because of it being dropped or banged, the complete charge may ignite and detonate. This is acceptable if a match is being applied to a cracker, because a bang is expected, but not when the person is igniting the 'blue touchpaper' of a skyrocket. The unsuspecting person is generally down on hands and knees to apply the match and is therefore in very close proximity to the danger. Many unnecessary injuries have occurred because simple rules have been disobeyed.

At the outset of my journey into rocket design and launching, safety precautions were one of the first procedures that I attempted to drill into my mind. I regarded the item that I was making as a bomb in disguise and had the greatest respect for it. In the early years, a few of my innovations had taken off, but in all different directions! The loud 'boom' and flying debris scared the very hair off my head and I realised at a tender age the inherent danger that is prevalent at all times. As I grew older, I started writing a safety manual and this was added to regularly, as a manual of this sort can never be complete! After each launch a post-mortem was held to discuss the various crews' input into the rocket's performance and, more importantly, any dangerous situations that may have existed.

From these discussions, many items in the safety manual were updated and others added. It was up to each crew leader to assess the operation that had been assigned to him and the safety code that was part of it. Any aspect on which there was dissension was immediately thrashed out, and on safety matters there was no compromise! Each of the group had an obligation to the safe discharge of the running of the range, be it for a static or flight firing. The very thought of an accident happening on the range was too horrendous to contemplate. Here is an extract from the safety manual as at the end of 1962:

- A barricade located some distance from the pad during any activity where a rocket motor is or is likely to be ignited must protect all personnel.
- Only remote operations may be carried out to a rocket that has fuel on board. Once the rocket is in the ready mode, nobody will be allowed to leave their position of safety. This will remain in force until the green flag is hoisted.

- The launch director will at all times have the safety key to the firing panel in his possession and it will only be inserted into the panel at his discretion.
- The countdown will commence on the advice of the range safety officer only.
- The rocket crew will take up their positions in the bunker after the red flag is hoisted. They will remain at their allocated posts until the range safety officer gives the 'All clear'.
- In the event of the rocket failing to ignite at the predetermined time a state of 'Unsafe' shall prevail on the range. All personnel will remain at their stations.
- Smoking is prohibited throughout the confines of the immediate launch area, that is an area of one hundred yards square, with the launch pad as the centre of this area.
- Members of the rocket crew shall wear protective clothing as deemed necessary by the launch director. Tin safety helmets will be worn at all times whilst working within the launch area.
- Members of the rocket crew will view the proceedings through the viewing ports only during the launch of the rocket.
- The launch director clad in an asbestos suit shall undertake fuelling of the rocket and members of the rocket crew may only then approach the rocket at the director's behest if in his opinion the situation is safe.
- All combustible materials must be separated and left in the open.
- When handling powder propellant, persons taking essential part in the mixing thereof will wear gas masks.
- Tracking crews will wear tin safety helmets and remain at their posts that are 1 000 ft in line with the pad and on either side of it. Only after the green flag has been hoisted may they return to the launch area.
- In the event of an upper stage returning to earth unfired, a waiting period of no less that three minutes shall be observed before the launch director only shall approach such stage. If it can be made safe, he shall raise his right arm and proceed to do so.

By following these simple guidelines, the SARRG had a perfect safety record and in addition a qualified first aider was always in attendance. To my knowledge the only untoward occurrence was when I had zinc dust blown into my eyes and a few eye-drops cleared that up.

The above safety procedures were taken from my file of schedules of events, where every detail of a firing were itemised and rigidly adhered to. The safety spec was part of this document. Each member of the various crews was issued with a new copy of this schedule at the pre-firing meeting. These meetings were generally held on the Wednesday before the coming weekend's activity. It was clearly understood by all members that any transgression of the safety regulations would mean instant dismissal from the group!

In no way do I wish to detract from the dangers involved in the pursuit of rocket science, but I strongly suggest that a thorough theoretical appraisal be made, not only into rocket design, but also into the requirements of the local laws pertaining to the experimentation with rockets. I know that in some countries it is banned totally because of the irresponsible manner in which individuals have approached the subject. Rockets are not toys and even the small models that are available commercially reach speeds in excess of 30 m/s! That makes them the fastest thing to be found in hobby shops.

The future

In the coming years great advances in rocket-propelled vehicles will be seen. At present there are designs for exotic motors. The hybrid rocket has already been tested and flown and has shown positive applications. It consists of a conventional-shaped combustion chamber with a solid fuel cast in it, and the liquid oxidiser is sprayed into the headend through injectors. This system obviates the fuel tank as well as the fuel turbopump. As yet the fuel capacity is the limiting factor, but the idea has merit in the weight-saving department.

Ion drive rockets have been built. The thrust is actuated by a beam of electrons travelling at the speed of light (300 000 km/s). This is the principle of the television tube in your home, and this device would only be usable in the vacuum of space because of the very low thrust levels obtainable. Also, the system for producing the electrical energy to produce the beam would perhaps be on the conventional nuclear reactor-boiler generator basis.

Preliminary calculations showed that for a beam giving a significant specific thrust for a reasonable time, a low current at very high voltage would be needed to produce 0,5 kg of thrust! The limitation of such a unit would be the efficiency of the generating system. The waste heat from the generating system would have to be radiated away and the size of the radiating equipment would be limited by weight restraints. It seems likely that units producing thrust in a few tens of pounds would be feasible. If a chemical rocket as we know it could loft one of these units into earth orbit, the continuous low thrust would eventually accelerate a space vehicle at tremendous velocity. Time to destination would be drastically cut compared to our present-day vehicles.

Nuclear energy has been considered for driving rocket-type vehicles. There appear to be three basic systems that could be employed.

- The product of fission itself could be converted to form thrust.
- Energy can be used to heat a working fluid that would be aimed and expanded through a rocket nozzle.
- Electrical energy can be produced, which in turn is used to produce ionised particles in a high velocity stream.

All the foregoing sound very attractive, but in each application the weight factor is of paramount importance! Nuclear devices are by their very nature heavy and mostly fairly large. Therefore a formidable launch vehicle of the chemical type would have to be used to place the probe/space vehicle in earth orbit. A nuclear rocket, even if it were possible to overcome all the weight problems now, could not be used to launch from earth owing to the radiation it would produce! Once in orbit, its chemical booster could propel the vehicle into a higher orbit where the radiation would be of little concern when the nuclear drive started.

The future indeed holds many possibilities for the exploration of our celestial neighbours and deep space. The propulsion systems to achieve these immense distances are on the drawing-board or in the static test stands at present. Consider the time span between the success at Kitty Hawk with the Wright brothers' first heavier than air aircraft and the landing on the moon by Armstrong and Aldrin. It stretched from 1903 to 1969 – just 66 years!

In my lifetime I have seen methods of travel leap from animal-drawn carts, early motor vehicles, aircraft that flew a little faster than the motor cars, commercial jet liners, to the huge space rockets of today! I consider myself extremely fortunate to have lived in a period where scientific advances in all fields have progressed at, may I say, an alarming rate. A look into the viable possibilities that the future holds is science fiction stuff at its best! The long term is space vehicles propelled by nuclear fusion. With this means a large vehicle could be propelled at incredible velocities by the force released from the fusion. Still further away is the ship powered by laser-driven lightsails utilising vast amounts of energy. These craft would have to be extremely light in weight, as the power required to propel them would be at enormous cost. It has been calculated that a manned spaceship to reach the nearest star would require a lightsail 960 km across! It would have to be assembled somewhere in the vicinity of Mercury, as it would require solar-powered lasers delivering the equivalent of thousands of times the world's output of power!

The exploration of space is a long-term project and since these types of spacecraft are already at the drawing-board stage, it can be seen that it is a very serious and real inevitability. The rockets of today are insignificant compared to those that are foreseen in the not-too-distant years ahead. Human beings have achieved a foothold on the unknown and are preparing themselves for the first visit to another planet. Mars is the most likely initial destination and the details for this trip are well advanced. Systems have been built or are in the process of being built to place people there, possibly within the first decade of this new millennium

Unimaginable as this may sound, the human being is a curious creature and will drive him- or herself to the very edge to obtain what is sought. The people that will bring these dreams to reality are perhaps in nursery school or primary school today and they will leap the gap just as our generation brought about the beginnings of the exploration of space. In time to come, people will look at our endeavours at rocket design and regard them as archaic and wonder how, in all that is beautiful, did they ever work? I stood in the Smithsonian National Air and Space Museum, in Washington, DC, and gazed in awe at a replica of Goddard's first successful liquid fuel rocket. I could not perceive that this was the actual beginning of the giant Saturn V moon rocket! But there it stood, all 3 m of it, two fuel tanks, a combustion chamber and the necessary plumbing. No aerodynamic shape, no covering around the working bits and pieces, just the bare essentials to heave the whole off the ground. How far had we come since that brilliant mind saw what no other could imagine. So in the grand scheme of things, 66 years was not a long time at all for a human being to leave the ground, fly in an ungainly machine and then prepare to reach into the heavens and touch the stars.

Chapter 12

The last chapter

In the intervening years since my last rocket roared into the air, I have not been unoccupied. Even before I was prohibited from firing rockets, I had developed a fascination for the birds of prey in our country.

They were, in my humble opinion, being needlessly persecuted and if this continued, we would be guilty of another crime against nature. It would be a disgrace if we as South Africans allowed a single species of wildlife to become extinct through our thoughtlessness.

With this mind, I set out to study these birds and come to an unbiased study as to their behaviour and biology. It took twenty years to gather the information and photographs of 56 different species, covering the southern portion of Africa from the Zambezi River down to the Cape Province. At the end of my peregrinations I considered myself to have enough information to write a book on these wonderful birds.

The book took two years to write and Purnell published it in 1975. The main thrusts were conservation and secondly, being illustrated with actual photographs, providing a ready, field identification guide. Now I was invited to lecture on ornithology!

I undertook to do a complete study on the African fish eagle and also the magnificent crowned eagle. This I gladly did, and the scientific papers were published. A grey cloud was lifted with the publication of the findings and the name of 'wanton killers' has largely been removed as a description pertaining to birds of prey. Certain quarters are now showing enlightenment! These two studies took 18 years to finalise, it was time well spent, and I enjoyed every minute of it.

I have been asked: 'How could you spend 40 years of your life studying these birds?'

My answer is very simple: 'What have you done in the last 40 years?' I am not trying to be smart, but I believe that every one of us has special talents and we should try and exploit these talents to the advantage of society in general. Only from this attitude may we all benefit.

The deep passion for rocket research has never left me and I have tried to keep up with the latest technology by reading and communicating with people involved in this field. If I look back over the years and have to summarise exactly what goals were met in South Africa during the period that I was actively engaged with rockets, I could say I, as an amateur, was responsible for:

- The first rocket to be fired for scientific purposes
- The first rocket to achieve a mile in altitude
- The first multi-stage rocket to be fired successfully
- The first to send a living creature in a rocket
- The only five-stage rocket to be successfully launched
- Sending a rocket to an unofficial altitude of 48 km
- Exceeding the speed of sound with a rocket
- The first liquid fuel rocket motor to be successfully run
- Firing the first liquid fuel rocket.

If these accomplishments are perceived as the result of amateurs, then they are indeed formidable. Also, bearing in mind that the amount of sponsorship was restricted to raw materials only, the picture becomes more remarkable. The efforts that were put into the entire programme by ordinary working men must be recognised as something special.

Every member of SARRG was dedicated to the final ambition that I foresaw all that time back. We had worked steadily in that direction and were seeing advances on our way to reach the goal. With each launch, another hurdle was overcome and our store of knowledge increased immeasurably. The rocket motor was fast becoming a reliable workhorse in our stable and we rapidly learned the secrets that it had in the beginning. But more importantly, we developed a respect for it that exceeded our understanding.

This was a piece of physics with which no liberties could be taken.

We had shown respect from the start, for to take risks was to invite disaster. On the odd occasion when I had got things wrong and a detonation occurred, it was most unpleasant to hear the bits of shrapnel sobbing over our heads. On exploding, the vast majority of flying debris appeared to travel horizontally and few went skywards. Bits and pieces were often found hundreds of metres away from the pad and we made an all-out effort to pick up all we could, in case some unsuspecting soul wandered onto a sharp piece of metal and received a nasty cut.

Because of the high thrust-to-weight ratio of the Saxon series of rocket, we were unable to make use of the weather-forecasting 'radio sonde' equipment. (It utilised a propeller-driven switching arrangement to select the various circuits, that is, barometric pressure, temperature and humidity.) The propeller was air-rotated by the slow and steady ascent of the balloon to which it was attached. If this apparatus were installed in one of the Saxons, the high speed would have ripped the propeller and its shaft to pieces. In view of this, we contrived to send maximum and minimum thermometers and aneroid barometers in recoverable capsules.

The thin metal bellows of the barometers had pieces of wax paper fixed in such a way that a thin needle scribed a mark on them. As the expansion of the bellows took place owing to the decrease in atmospheric pressure, the geared mechanism moved the needle. Remember, the greater the altitude, the less the pressure, the greater the expansion of the bellows. On recovery, the instrument was placed in a large jar with an aircraft altimeter and the air evacuated. The height was estimated when the needle on the wax paper reached the end of the scribed mark, and this was compared with the altimeter reading. Crude perhaps, but it gave a reasonable estimation.

A clockwork timer, the type that was used for delaying the shutter in photography, operated the ejection system for the capsule and parachute. The parachute was of such dimensions that it created sufficient drag to slow the package's descent, but not allow it to drift. If the latter was to happen, it would be anybody's guess as to where it would land, and dropping from a considerable altitude, it could have come down kilometres away.

As I mentioned before, to maintain a stable vertical trajectory, the centre of aerodynamic pressure must always be behind the centre of gravity of the rocket. This meant that for the various payloads that were carried, different size fins were necessary to meet this requirement. The easiest way to accomplish this was by increasing or decreasing the span of the fin planform, in other words, the base of the fin that rested on the launch pad. By so doing, I could gain or lose a small amount of weight of the rocket, to place the two centres in the correct positions.

To calculate the centre of aerodynamic pressure, such dimensions as diameter, length, number of fins, transitions in body diameter and nose shape of the rocket are the basic requirements to equate. This calculation used to be a long bit of maths, using a slide-rule. It was made easier with the advent of the calculator and now I have the whole package loaded into my computer. I call up the program, feed in the data, press the 'Enter' key and there is the answer. Not five minutes. The only manual work requires finding the exact point of the centre of gravity of the ready-to-go rocket. The ideal position for the CP is two body diameters behind the CG. In rocket language we refer to it as being two calibres behind the CG – the above two positions refer to liquid and solid fuel vehicles to retain stability in vertical flight.

The fin placement has been touched upon, but the need for accurate alignment and fixing is a major factor in the stability of the rocket's performance. I had always preferred four fins to three, albeit that the latter save weight and have been successfully flown. I found that in the building of the large Saxon series, it was easier to align the four-fin configuration at 90° apart than to align three fins at 120°. Experimental rockets were flown with six fins spaced 60° apart and were discarded as no significant improvement in stability was noted as well as the disadvantage of the extra weight carried.

I had attempted to design and fly rockets from an early age as an attempt to copy what, at that time was pure science fiction. But with the passing years and an unsatiable thirst for knowledge in the science of rocketry, I soon came to the realisation that what I was pursuing was no longer fiction but something that was real and in its infancy throughout the world. It was a brand-new science and those that were involved in it were only too keen to learn from others who had the same dreams and aspirations. This exchange of ideas was very soon to change and rockets and rocket propulsion were suddenly a no-no.

During those early years, movie news from the USA showed a little of what was being attempted. I remember seeing quite a few impressive blow-ups and these stuck in my mind, reminding me that if I continued in my quest, I should expect the same. I had no one to teach or show me the intricacies of rocket motor design and it was a frustrating period in my life as I watched my efforts explode or simply stand dead-still, smoke and fire spluttering from the tail. It was not until I grasped the basic fundamental that all the action must take place *inside* the combustion chamber and not give an impressive display of sparks in the atmosphere that my long trial ended.

From that moment on, the rocket became an object that I could plan correctly on paper and watch it work in the field, with fewer nerve-shattering bangs. There is no way to describe the inner feeling I got, watching a rocket that I had designed streak off the pad and head into the sky, just as the numbers showed it would. With each launch something new was learnt, but the greatest lessons came from those that failed to perform, such as the unreliable ignition system for multi-stage rockets and the resulting solution that meant that better and bigger rockets could be designed.

In this brief recounting of my experiences in this science, I have presented the events as they transpired, reading through old diaries and newspaper clippings to ensure accuracy and thoroughness. Remember, this all took place more than 30 years ago, and the genesis had sparked into life two years after World War II had ended. It was a long time ago, but the spark is still there prompting me in my sleep and at all odd times during the day. It was and is something that was part of my very being, and as such, time cannot dim the grand memories. It is a pity that I did not realise my ambition, but I still get an inexplicable feeling when a person of my age is introduced to me and says: 'Aren't you the Des Prout-Jones that flew rockets and sent a mouse up in one?'

Epilogue

The fire is no more

The sands of time have run quickly and I see before me now, nearly 30 years later, the faces of the men of the SARRG. All but two, to my knowledge, have passed on, for they were grown men and fathers, while I was yet a boy. They shared my enthusiasm and disappointments and they never queried my judgement, although they were my best critics. To them, I give my most heartfelt gratitude, for without their skills I could never have achieved what I had set out to do.

A heart attack claimed Jack Holloway and last year I heard that Gordon was not well and I hastened to be with him. I found him in a chair at home, his left leg pointed stiffly on the floor. He still wore the scowl that I had come to know so well and his wife told me that he had had both kneecaps replaced. The right one was fine, but the left was infected. The poison had invaded his body and a huge abscess had formed in his right side. This required continuous draining and appeared not to heal. I visited him on Monday mornings for some months and we spoke about many things. If the sun was shining I would push his chair outside and we would talk and drink tea in the warmth. When he left us, I felt sorrow but at the same time I knew that he was out of pain and riding where we had sent the last rocket.

I had, unfortunately, lost touch with most of the others and I received a surprise one day when the receptionist paged me to say I had a visitor.

It turned out to be one of my tracking men. He had heard that I was working for a large electronic company so, being in the area, he dropped in to say 'Hallo'. I had not seen this member of the tracking crew for some 15 years and it was great to sit and recall the times and events that we had shared. I learnt that his wife had passed on a few years earlier and that he was staying with one of his sons. We met again shortly after this meeting but soon after I lost contact with him. I have not heard of him since.

The Chief Inspector of Explosives, I heard, had retired and I did not care to find out who had filled his position. I think I have one letter from his department still in my possession. When I came across it, a myriad of memories flashed through my mind. Strange, but the most vivid recollection I have of this man is the time he bawled me out at our very first meeting! I really have to put on the thinking cap to bring to mind some of the many meetings I had with him. They have all seemed to merge into the distant past, like a mist being swept away slowly as the sun rises in the morning.

If I close my eyes, I can see the faces of all the members of the group, hear their voices and their own brand of humour. Their dismay as I handed out tin safety helmets to be worn at all times on the range! I am not going to relate the name they had for to these helmets! They were a noble bunch and their enthusiasm for rocket science was matched only by my own.

In 1998 I visited the Kennedy Space Center in Florida, USA, and the flickering began once more! I have returned there on two more occasions and now I cannot get the hook out of my mouth! Although I have not been involved in any way with rockets for the past 36 years, these visits ignited the flame once more. Not on the grand scale as before, but now I have grandsons that share this obsession and are only too excited to learn and take part in model rocketry. Small rocket motors are commercially available, as are rocket kits for young aspiring rocketeers to assemble and fly with no danger.

My grandsons and I go through the mathematics and build our own rockets, using these small motors and spend a great Sunday afternoon launching them. To hear the squeals of delight from my granddaughters, who act as firing officers, when the rockets lift off makes it all worthwhile!

On the Internet, I learnt that South Africa had entered into collaboration with Israel to establish a rocket research centre, but had ceased activities in the late nineties. It was sited on the West Coast, facing the Atlantic Ocean. The rocket that was being researched was called the Jericho, but nothing else was disclosed. No reports of any launchings or progress of research: it was all clouded in the veils of secrecy. It sings of military applications rather than space involvement. Sad to think that the exploration of space could be undertaken here, just as the European Space Agency has taken off and is in competition with Nasa with regard to satellite orbiting today.

When one thinks back to the Apollo programme to land a man on the moon and bring him back safely, it created work for nearly 300 000 people in 20 000 factories throughout the USA. It was a tremendous boost for the economy and dropped the unemployment figures drastically. I am not for one moment suggesting that we do things on the same scale, but apart from adding to the exploration of the universe, we would also be providing jobs for many thousands of unemployed. The challenge has been there for more than 40 years but the baton has never been picked up!

Thinking back to that historic day in July 1969 when the first human stepped onto the moon, the entire population of the Earth were united as one. All nations, whatever their colour, religion or political beliefs, were bonded into one huge emotional state as Neil Armstrong uttered the immortal words: 'One small step for man, one giant leap for Mankind.' That state of oneness has not been experienced since and all too soon the journey to the moon took on the 'Ho, hum' attitude and the focus turned back to the material and plastic world. The courage of those first space travellers has been forgotten and buried in the archives of history.

Today the grass and scrub remain unchanged on our firing range, as they were all those years ago and the concrete block is overgrown with weeds, making it difficult to find. Even more difficult to imagine is that the launch tower once stood as a silent sentinel, watching over the murmuring veld. The same rusted lock secures the heavy entrance gate that once gave access to the sand road leading to the bunker. Plovers still scream their objections to anyone venturing onto their breeding domain, and gamebirds rise with a flurry of wing beats, breaking the tranquillity. A slight ridge in the flat savannah is the only indication as to where the bunker was. In it we had sought shelter in those exciting days of rockets research and, if one listens carefully, Gordon's voice may be heard intoning the countdown. The passage of time has all but eradicated our smal beginnings, but the results are burnt into the annals of this country – and they will never pass into oblivion as long as there are people who hunger for knowledge.

References

Anon, 1899. *Halfhours in air and sky*. James Nisbet & Co.
Clarke, A C. 1957. *The making of a moon*. Frederick Muller Ltd. London.
Collins, M. 1974. *Carrying the fire*. Farrar Straus & Giroux. USA.
Chaiken, A. 1994. *A man on the moon*. Harmondsworth: Penguin.
Chaiken, A. 1997. *Air and space*. The Smithsonian Institution. Bullfinch Press. USA.
Fowler, E. 1999. *One small step*. Smithmark. New York.
Furniss, T. 1998. *One giant leap*. Carlton. UK.
Gatland, K & Kunesch, A. 1953. *Space travel*. Allan Wingate. London.
Heppenheimer, T A. 1997. *Countdown: a history of space flight*. John Wiley & Sons. New York.
Humphries, J. 1956. *Rockets and guided missiles*. Ernest Benn. London.
Lee, W. 2000. *To rise from earth*. Cassell & Co. London
Neal, V, C S Lewis & F H Winter. 1995. *Spaceflight*. Smithsonian Guides. Macmillan. USA.
Rosen, M W. 1954. *The Viking rocket story*. Faber & Faber. London.
Stine, G H. 1994. *Handbook of model rocketry*. John Wiley & Sons. New York.
Sutton, G P. 1956. *Rocket propulsion elements*. John Wiley & Sons. New York.